Advanced Sciences and Technologies for Security Applications

Indexed by SCOPUS

The series Advanced Sciences and Technologies for Security Applications comprises interdisciplinary research covering the theory, foundations and domain-specific topics pertaining to security. Publications within the series are peer-reviewed monographs and edited works in the areas of:

- biological and chemical threat recognition and detection (e.g., biosensors, aerosols, forensics)
- crisis and disaster management
- terrorism
- cyber security and secure information systems (e.g., encryption, optical and photonic systems)
- traditional and non-traditional security
- energy, food and resource security
- economic security and securitization (including associated infrastructures)
- transnational crime
- human security and health security
- social, political and psychological aspects of security
- recognition and identification (e.g., optical imaging, biometrics, authentication and verification)
- smart surveillance systems
- applications of theoretical frameworks and methodologies (e.g., grounded theory, complexity, network sciences, modelling and simulation)

Together, the high-quality contributions to this series provide a cross-disciplinary overview of forefront research endeavours aiming to make the world a safer place.

The editors encourage prospective authors to correspond with them in advance of submitting a manuscript. Submission of manuscripts should be made to the Editor-in-Chief or one of the Editors.

Carl S. Young

Cybercomplexity

A Macroscopic View of Cybersecurity Risk

Springer

Carl S. Young
New York, NY, USA

ISSN 1613-5113 ISSN 2363-9466 (electronic)
Advanced Sciences and Technologies for Security Applications
ISBN 978-3-031-06996-3 ISBN 978-3-031-06994-9 (eBook)
https://doi.org/10.1007/978-3-031-06994-9

This Springer imprint is published by the registered company Springer Nature Switzerland AG
The registered company address is: Gewerbestrasse 11, 6330 Cham, Switzerland

Only entropy comes easy.

Anton Chekov

To Geraldine Prose Young, MD
September 30, 1926–June 14, 2020

Foreword

In 1996, I was selected to start one of the first "Computer Crime Squads" in the New York office of the FBI. At that time, the Internet was starting to become a common utility, where access was facilitated by dial-up connections from telephone landlines.

As an investigative agency, the focus of the FBI was to understand how the Internet would affect investigations and what evidence would be available whenever this new medium was utilized. As I jumped into what was then a new and fast-growing field, other experienced FBI agents told me the one person with whom I should consult was Carl Young.

In those days, Carl was heading a unit within the Engineering Research Facility at the FBI Academy in Quantico, VA. I scheduled a meeting with Carl in his office, and during our discussion, it became clear why I was directed to speak to this individual.

Carl eventually became a member of the Senior Executive Service within the FBI Intelligence Division. Despite his senior status, Carl sought to understand "ground truth" by listening to the operational requirements of fellow investigators. He leveraged his education in physics to solve problems that had a significant impact on national security. He possessed a rare combination of operational experience, academic training, and facility for problem solving, and his input was valued by executive management, scientists, and FBI agents in the field alike.

That initial meeting marked the beginning of a relationship that has continued for over 20 years. Carl and I retired from the FBI in 2000, and each of us transitioned to the private sector. Carl joined Goldman Sachs in New York and I founded Stroz Friedberg, LLC, one of the first computer forensic firms.

We continued to discuss cybersecurity risk, his experiences at Goldman, and my experience working hundreds of engagements for clients and ultimately coalesced into a shared vision of what was needed in the marketplace. I invited Carl to join my firm and he did, launching what he aptly called the "Security Science" division.

We were immediately able to draw from a unique set of experiences and internal case studies our company owned. Those studies enabled us to review "postmortems" in an attempt to identify areas of commonality. We repeatedly observed the same types of issues across organizations and agreed their root cause was viewing cybersecurity as an exclusively technical issue. This realization continues to influence our thinking on cybersecurity risk management.

In Carl's fifth reference book on security risk management, *Cybercomplexity*, he addresses one of the most challenging issues in cybersecurity. He does so by leveraging elementary probability and information theory to develop a simple model of complexity in IT environments while drawing on analogies from physical science. He also reveals why specific types of security controls are required to reduce complexity and thereby address cybersecurity risk on an enterprise scale.

Cybercomplexity does not indulge in technical jargon or "acronymology" and is therefore accessible to non-scientists. It represents another example of this author's success in applying science to security and is a reaffirmation of the close connection between the two areas. I have heard Carl say he hopes he can at least modestly improve the "signal-to-noise ratio" of security risk management. This book has clearly done that and more.

For any individual wishing to understand the foundations of cybersecurity risk, this book offers a resource to be repeatedly consulted. It offers unique insights into issues that affect all IT environments and that have confounded cybersecurity professionals for years. Importantly, it can enhance the sophistication of reporting to senior executives and boards of directors with respect to cybersecurity risk management challenges and best practices.

The level of success we achieve in cybersecurity is greatly affected by how we get to the core of its problems. By embracing this book's lessons, practitioners, managers, and executives will promote a security culture that reflects the thoughtfulness of a scientific approach.

Edward M. Stroz
Founder, Stroz Friedberg LLC
New York, NY, USA

Preface

What does "cybersecurity" actually mean? A book's title should reflect its content as a matter of principle if not professional courtesy. Specifically, is the use of "cyber" appropriate in this context noting the term is already a fixture in the English vernacular? To answer this question, it is helpful to explain its origins.

The first use of the term cyber was in Norbert Weiner's famous work, *Cybernetics.*[1] It derived from the Greek word *kubernetes* or "steersman" and has the same root as the English word "governor" as in the controller or speed limiter of a machine. The term cybernetics is a reference to the confluence of communication and control, which is the central theme of *Cybernetics.*[2] Weiner, a professor of mathematics at the Massachusetts Institute of Technology, coined the term in 1948.

Weiner recognized that "the problems of communication and control engineering were inseparable."[3] For example, the otherwise pedestrian task of eating with a fork is less appreciated as a problem in communication and control. The brain is iteratively processing positional data while updating the signals transmitted to the muscles that guide the fork to its destination. When viewed in this light, shooting down a missile and eating with a dining utensil are each exercises in optimal prediction.

Weiner made the profound revelation that the central focus in addressing such problems should be the signal message, which is characterized as "a discrete or continuous sequence of measurable events distributed in time."[4] Statisticians refer to these sequences as *time series.*

Moreover, he reasoned the solution to problems in optimal prediction could be found in time series statistics, and more specifically, in finding an explicit expression for the so-called mean square error of prediction. The presence of background noise

[1] N. Weiner, Cybernetics; Or Control and Communication in the Animal and the Machine; MIT Press, John Wiley and Sons, New York, NY, 1948.

[2] A. Broadhurst and D. Darnell, An Introduction to Cybernetics and Information Theory, *Quarterly Journal of Speech*, Volume 51, 1965.

[3] N. Weiner, op. cit. 1948.

[4] Ibid.

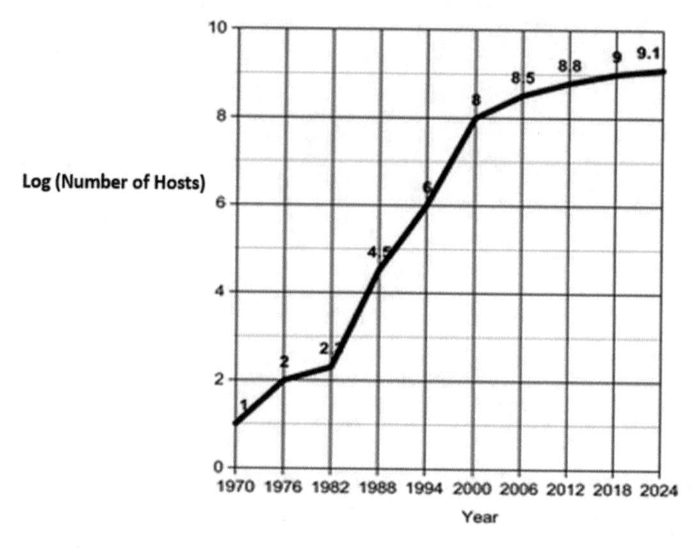

Fig. 1 Internet usage by year

is the complicating feature, and recovering the transmitted message depends on the statistical nature of the message and the corrupting noise.

Weiner notably posited that humans and machines were equivalent with respect to communication when viewed through a cybernetic lens. In that vein, cybernetics and cybersecurity share a common theme in that the predicate for both disciplines derives from machines as communication devices.

This thematic connection is particularly significant in light of the ascent of computer technology in communication. These days, computers (broadly defined) are integral to most forms of communication, and reliable access to the Internet has become a necessity.

To gain some perspective, Google alone processes more than 40,000 searches *per second*.[5] One 2021 estimate of the number of Internet-connected devices is 27.1 billion.[6] Figure 1 illustrates the explosive adoption of the Internet as a means of communication, noting the vertical axis is a logarithmic scale.[7]

Although Weiner was among the first to recognize the inherent relationship between machines and communication, even he might not have anticipated the rapid evolution of modern computing. The ubiquitous presence of information technology is in part a result of improvements in electronic storage, channel capacity, and connectivity. Techniques for mass production have also evolved so that information technology is accessible to broad segments of the population.

---

[5] www.forbes.com; May 21, 2018.

[6] www.cisco.com.

[7] Internet Count History, Internet Systems Consortium.

However, machines and humans perhaps have an even deeper relationship through the numerous applications that can be downloaded via the Internet. Nowadays, individuals possess a digital identity defined by these applications. Furthermore, digital identities are arguably displacing physical identities as online activities increasingly substitute for personal interactions. The COVID-19 pandemic might have accelerated this phenomenon, but continued virtualization is inevitable.

Cybersecurity is clearly affected by the trajectory of information technology and its usage. However, although the physical and virtual worlds increasingly overlap, they remain distinct in significant ways. For example, the methods used to restrict physical access, e.g., sensory perception, locks, physical barriers, and alarms, are not applicable to the virtual world. Less intuitive methods must be deployed to restrict electronic access. This problem is compounded by the perpetual desire for convenience, which has potentially disastrous consequences when sharing information online.

Increased computational power as predicted by Moore's Law resulted in miniaturization that has accelerated the dependence on information sharing.[8] Mobile devices and the accompanying fluidity of network boundaries have ramped up expectations of convenience while amplifying the need for cybersecurity. Smart phones, tablets, etc., enable unprecedented access to information irrespective of physical location.

In fact, the very notion of a physical boundary has become increasingly fuzzy in the virtual world. In many scenarios, it is downright meaningless. Network boundaries can be generously described as fluid, where Wi-Fi and cellular technologies extend the perimeter to anywhere within range of a radiating access point or cell tower. Convenience on that scale is inevitably accompanied by an increased potential for information compromise.

Note the nature of electronic information itself has security implications. Specifically, although information might have been stolen, it might not actually be missing. In other words, physical access to an item of value is not required in order to steal it.

One strongly suspects something is awry with respect to traditional approaches to cybersecurity risk management given the legacy of successful cyberattacks. Such attacks persist despite countless regulatory requirements, security policies, security technology standards, and sophisticated security technologies. Significantly, data breach *postmortems* frequently point to the same *modus operandi*.

One plausible explanation for the current situation is cybersecurity continues to be viewed as a technical issue rather than as a traditional problem in risk management. Therefore, the focus is on technology fixes simply because technology facilitates information exchange. The fact is that the root causes of cybersecurity risk often have nothing to do with technology.

According to cybernetics, man and machine are similar in how they process information. In cybersecurity, it is the interaction between man and machine that is most significant. To be clear, networked computers do precisely what they are designed to do: enable information sharing. Unfortunately, information security and information

[8] Moore's Law, named after Gordon Moore, CEO and co-founder of Intel, states that the number of transistors in an integrated circuit doubles every two years.

sharing are inherently in tension, which explains why cybersecurity professionals perpetually face an uphill battle.

At the risk of stating the obvious, cybersecurity would be much less challenging without the Internet. Bad things can occur when billions of invisible individuals exchange information via a highly distributed and massively convenient network. But Internet access is synonymous with easy information sharing, which is now integral to our personal and professional lives.

The Internet differs from other electronic networks in that the network nodes need not be physically connected. This property also has significant security implications. Most notably, it frees both authorized network users and miscreant network attackers from being constrained to specific physical locations.

Consider the telephone network before the days of IP telephony. Previous technology limited the power and flexibility of POTS devices but also reduced the potential for information compromise because physical access to the equipment was required.

I recall my university's more mischievous students perpetrating hacks against "Ma Bell" as the phone company was affectionately called. They were forced to physically access the equipment in order to commit their prank and thereby thumb their noses at "The Establishment," which the phone company personified in the nineteen seventies.

Modern hackers are both ethically less benign and physically less constrained than their forebears. Nowadays, a telephone is merely another device on the IP network, and therefore, hackers need not leave the comfort of their homes to wreak havoc on organizations and individuals alike.

There are other aspects of electronic networks that affect the security risk profile. Network communicators are invisible, and messages can be routed to their destination without attribution. Malicious actors exploiting technology vulnerabilities and/or human foibles drive the requirement for a strategy of "zero trust" when attempting to access information assets.

In the end, cybersecurity is about securing electronic information that is processed by machines, operated by humans, and connected via networks. These networks are vast in scope, opaque in detail, and highly diversified. The result is a multifaceted environment that enables unprecedented information sharing but is also ripe for exploitation. The proliferation of vulnerabilities in such environments is almost inevitable.

Security risk assessment outcomes are affected by how one addresses such vulnerabilities, and high-severity examples clearly require addressing in a timely manner. However, the aggregate effect of risk factors also impacts the potential for information compromise that includes non-technical issues associated with processes and workflows. Such effects are generally not visible unless IT environments are viewed through a sufficiently broad lens.

To that end, this text explores the *macroscopic* forces that affect IT environments on an enterprise scale and the implications to cybersecurity risk management. Specifically, *Cybercomplexity* is divided into four parts: (1) Security Risk Fundamentals, (2) Stochastic Security Risk Management, (3) Enterprise Cybersecurity Risk, and

(4) Cybercomplexity Genesis and Management. The following paragraphs describe the individual chapters within each of these sections.

Chapter 1, "Core Concepts," discusses the conceptual foundations of security risk and risk assessments. Although many readers may already be familiar with many of these concepts, thinking rigorously about risk requires a grasp of the basics, starting with the definitions of threat and risk.

Chapter 2, "Representing Cybersecurity Risk," focuses on the representation of risk-relevant phenomena. The objective is to explain concepts essential to understanding and conveying risk-relevant information.

Chapter 3, "Scale and Scaling Relations," represents a continuation of Chap. 2, where the focus is on describing relationships between risk-relevant parameters. A key result of the theory of complexity in IT environments is that the perspective or "scale" used to assess cybersecurity risk affects the assessment results. In particular, the existence of linear versus nonlinear scaling relations can have significant operational implications.

Chapter 4 is entitled "IT Environment Dimensions and Risk Factors." It describes a multi-dimensional representation of IT environments. These dimensions encompass the sources of risk factors for information compromise. The number of risk factors across all dimensions impacts cybersecurity risk on an enterprise scale and drives the requirement for a macroscopic approach to cybersecurity risk management.

Chapter 5, "Security Risk Management Statistics," begins the second section of the text. This chapter provides the conceptual foundations for a statistical description of IT environments. This description requires genericizing risk factors and security controls, where risk factors are either managed or not according to a binomial probability distribution. The result is IT environment states consisting of unique combinations of managed and unmanaged risk factors, thereby paving the way for applying the information theoretic formalism that follows next.

"Information Entropy" is the title of Chap. 6. Entropy is a concept derived from information theory, and it is fundamental to the model of complexity in IT environments. Specifically, information or Shannon entropy quantifies the uncertainty of a probability distribution, and it is the probability distribution of security risk management outcomes that leads to an expression for the unpredictability of IT environment states introduced in Chap. 5.

Chapter 7, "Complexity and Cybercomplexity," begins the third section of the text, which is entitled, "Enterprise Cybersecurity Risk." This chapter defines the general notion of complexity in terms of unpredictability and applies a binary stochastic security risk management model to IT environments. The result is a scaling relation for IT environment complexity in terms of the number of probable states of managed and unmanaged risk factors. The unpredictability of those states describes complexity in this context.

Chapter 8, "Cybercomplexity Metrics," specifies metrics that arise from a stochastic security risk management process. Although these metrics do not enable security control calibration, they represent a first step toward quantifying the effects of complexity in IT environments. Perhaps more significantly, they highlight the relevance of scale in assessing cybersecurity risk as well as substantiate the requirement for the macroscopic security controls discussed in Chap. 10.

Chapter 9 "Cybercomplexity Root Causes," begins the final section of the text, "Cybercomplexity Genesis and Management." This section identifies the origins of complexity in IT environments and specifies requirements for its management. Chapter 9 delineates the most prominent root causes of Cybercomplexity, which are the progenitors of many cybersecurity incidents. This chapter is arguably the most operationally consequential in this section. Identifying and addressing the root causes of Cybercomplexity are necessary in reducing the potential for information compromise.

Chapter 10 "Macroscopic Security Controls," specifies the security controls that have a systemic effect on cybersecurity risk management. As their name implies, these controls function macroscopically, i.e., on an enterprise scale, and are antidotes to the root causes identified in Chap. 9.

Chapter 11 "Trust and Identity Authentication," focuses on trust in identity authentication, which is an issue that is currently top-of-mind in cybersecurity risk management. The concept of "zero trust" is particularly in focus. This chapter discusses how trust in identity can be formalized via a stochastic formulation of identity authentication.

Chapter 12, "Operational Implications," is the final chapter of the book. As its name implies, it focuses on the operational implications of cybercomplexity. Although such implications have been identified throughout the book, this chapter discusses the key implications in more detail as well as presents them in one place for reference. Candidly, these implications are mostly common sense and fortunately tend to confirm intuition about cybersecurity risk. Nevertheless, common sense is not necessarily common, and the implications can both inform and enhance traditional assessments of cybersecurity risk.

Finally, the principal focus of *Cybercomplexity* is on characterizing cybersecurity risk on an enterprise scale. The Cybercomplexity model is admittedly based on an idealized form of cybersecurity risk management. The contention is a probabilistic approach is helpful if not required to simplify IT environments and thereby examine cybersecurity risk at the desired scale.

The breadth and variability of IT environments have historically undermined such efforts. The intent is to overcome these obstacles by making simplifying assumptions in the hope of generalizing the results to more realistic scenarios. The good news is the lessons so derived make sense, and their broader applicability seems reasonable if not compelling.

New York, USA Carl S. Young

Acknowledgements

Writing acknowledgments can be challenging since the correlation between past contributions and present achievements is not always clear. Fortunately, the contributions of each individual mentioned below are timeless, which has made writing this particular acknowledgment especially easy.

My late parents, Dr. Irving Young and Dr. Geraldine Prose Young, have contributed significantly to everything I have achieved. Their enthusiastic support for their offspring has inspired family members across generations. My sisters, Diane Uniman and Nancy Young, continue the tradition by demonstrating love for me and support for my work.

Certain family friends deserve special mention due to the longevity and intensity of our friendship. Bill Seltzer, Vivian Seltzer, and Sora Landes have been like parents to me. Fortunately for them they have been spared that unenviable burden. I am grateful for their ongoing encouragement on this particular project as well the love they have shown me and my family over many decades.

I am fortunate to be close to a number of childhood friends who are still a key part of my life. These friends include Fran Davis, Maggie Degasperi, Dave Maass, Lisa Maass, Peter Rocheleau, Ruth Steinberg, and Jim Weinstein. The duration of these relationships exceeds 300 person-years, which is a testament to each individual's endurance and genuine affection.

I would be remiss if I did not mention the New Yorkers who are as close in spirit as they are in proximity. They include Maurice Edelson, Dick Garwin, Bob Grubert, Mal Ruderman, Paula Ruderman, and my cousin, Kate Smith. Donna Gill, collaborative pianist and vocal coach extraordinaire, deserves special mention, recalling our times together on the Upper West Side of Manhattan and *both* porches of her apartment in Chautauqua, NY.

Ed Stroz, the founder of Stroz Friedberg LLC, and more recently my partner in Consilience 360, has been a close friend and ally since I moved to New York City in 2000. His pioneering efforts in computer forensics have been influential in the evolution of computer security as well as my own thinking on that topic.

Steve Doty, the founder of Defensible Technology, is a friend, colleague, and a fellow advocate for risk-based security. It is rare to find a cybersecurity practitioner who can reason about risk from first principles as well as design a secure network. Our numerous discussions on cybersecurity risk management have helped me bridge the gap between theory and practice.

As a professional statistical mechanic, Chris Briscoe possesses a physicist's understanding of both forms of entropy. Our discussions have been invaluable in helping me appreciate both the benefits and limitations of a probabilistic approach to security risk management.

Finally, I must acknowledge and thank my colleagues in the IT Department at The Juilliard School, with whom I have had the privilege of working since 2016. I have not worked with a more competent and dedicated group of technology professionals. My pride in their accomplishments in a challenging environment is only exceeded by my gratitude for all they have taught me.

New York, NY, USA Carl S. Young

Introduction

Cybersecurity professionals manage the risks associated with the threat of information compromise and information-related business disruption. The simplicity of their job description belies the difficulty of their job. Information technologies are designed to make information sharing easy, which is potentially antithetical to the security risk management mission. Furthermore, network users crave convenience and, therefore, are motivated to circumvent security risk management methods, also known as security controls in the security vernacular.

People plus information technology is a recipe for information compromise. Software and hardware configured to work in harmony and perform at scale will inevitably suffer from flaws that are exploited by individuals with varying agendas and a lot of time on their hands. In addition, network users frequently behave in ways that make the attacker's job easier. Exploiting such behavior is the basis for certain attacks, most notably social engineering.

The specter of relentless attacks, no shortage of attackers, and the prominent role of technology in information management compel cybersecurity professionals to concentrate on addressing vulnerabilities via technical solutions. Although an exclusive focus on such vulnerabilities might be operationally expedient, a comprehensive strategy must include a more expansive view of cybersecurity risk.

Unfortunately, cybersecurity professionals often face a difficult choice due to time and resource constraints. Tactical issues become priorities because the clock begins ticking immediately after vulnerabilities are published. In addition, a restrictive cybersecurity strategy can place security professionals on a collision course with business types who have their own obligations and constraints.

There also appears to be a pedagogical bias in favor of tactical security measures. Many books have been written on specific attacks and vulnerabilities, yet surprisingly little has been published about the actual root causes of cybersecurity incidents. Perhaps tackling these root causes is considered too difficult or not in a Chief Information Security Officer's (CISO) purview. Whatever the reason, the absence of pedagogy with respect to the drivers of cybersecurity risk is conspicuous.

A modern IT environment consists of multiple technologies that support numerous network users. Such environments are routinely if informally described as *complex*, where such a condition is generally acknowledged to be a significant contributor to the potential for information compromise. Despite this acknowledgment, a model of complexity in this context has proven elusive.

Any realistic model of cybersecurity risk must include a feature that at least resembles complexity. The principal obstacle to date has likely been the variability of IT environments. Variability resists generalization, and a model is by definition a generalization. Moreover, the variability is not uniform. It changes as the scale of the IT environment expands, which is a foreshadowing of future discussions.

An enterprise-level perspective is required to characterize complexity in IT environments because the latter is a singularly macroscopic phenomenon. However, such a perspective is not possible if the focus is exclusively on individual devices or network elements. A desired outcome of a macroscopic view is the identification of a *scaling relation*, i.e., a quantitative relationship between risk-relevant parameters. The quest for this outcome is ultimately the impetus for adopting a probabilistic, i.e., *stochastic*, model of cybersecurity risk management.

Other risk management disciplines estimate the probability of a future threat incident based on a probability distribution of historical incidents. Why can't the same procedure be used to assess cybersecurity risk? For example, medical science is able to estimate the probability of contracting various diseases based on the presence of one or more "risk factors." The unambiguous correlation between lung cancer and smoking tobacco is a prominent example.

One significant difference between diseases and information compromises is that humans are physiologically similar or similar enough. Therefore, any smoker is a potential candidate for inclusion in a sample drawn from the general population. Numerous individuals have unwittingly provided a statistically significant sample by smoking for differing numbers of years. The relatively consistent anatomy and physiology of humans enable generalizations of sample results to the remainder of the population.

The figure below shows the relationship between the duration of a smoking habit and the cumulative risk, i.e., the probability as a function of time, of developing lung cancer.[9] These curves derive from the unfortunate lifestyle choices made by members of the sample population.

[9] https://www.ncbi.nlm.nih.gov/books/NBK1554/figure/A193/; The figure is redrawn from Peto, R., Darby, S., Deo, H., Silcocks, P., Whitley, E. & Doll, R., Smoking, smoking cessation, and lung cancer in the US since 1950: Combination of national statistics with two case—control studies. Br. med. J., 321, pp. 323–329, 2000.

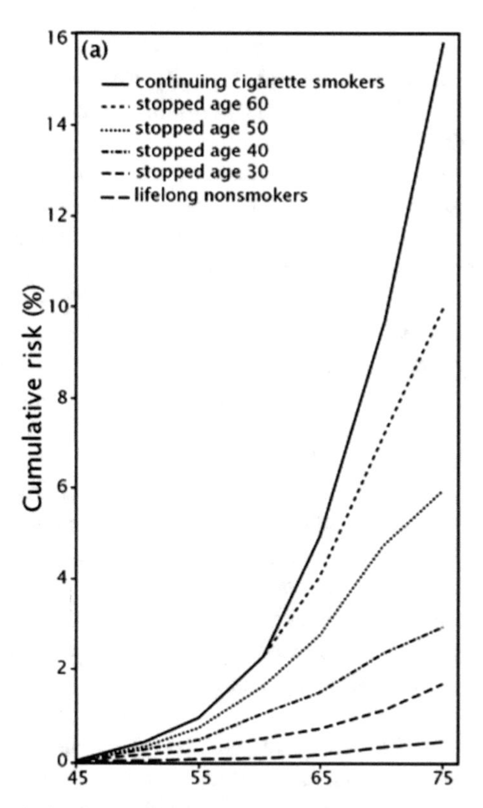

Cumulative risk of developing lung cancer as a function of the number of years smoking tobacco

There are several reasons why this type of data is not available in IT environment threat scenarios, where cyberattacks are the equivalent of diseases in the medical world.

First, IT environments are dynamic. Computer users are constantly beginning and ending online sessions, visiting Web sites, and accessing disparate networks via Wi-Fi connections, etc. Network user profiles will also vary depending on the individual's role within the organization. Network, workstation, and application updates occur less often but are not infrequent.

Therefore, comparisons of cybersecurity threat scenarios are difficult since risk factors for information compromise fluctuate over relatively short timescales, and even small fluctuations can significantly affect the security risk profile of an IT environment. In contrast, humans do not change appreciably over timescales relevant to most medical studies. Large population studies that require months or even years to conduct can be completed without concern that the sample will change appreciably and therefore render the results irrelevant.

Animals can sometimes be used as proxies for humans in medical experiments, and the results of these experiments are sometimes generalizable to humans with appropriate caveats. Despite the prevalence of “mice” in IT environments, animal studies are not possible in assessing cybersecurity risk. As an aside, *Homo sapiens* owe a huge debt of gratitude to their earthly co-inhabitants for sacrifices made in the name of human health.

Note IT environments can vary considerably when viewed at close range. Even relatively small IT environments contain numerous, diverse elements when examined in detail. IT environments are similar when viewed at a high level as evidenced by the OSI Model. But a more granular view reveals considerable variation.

Second, it is difficult to conduct experiments to quantify the effect of cybersecurity controls. For example, imagine conducting double blind experiments to test the efficacy of a security control in the same way vaccines for SARS-Cov-2 have been tested. Moreover, isolating the contribution of a single IT environment element to the overall potential for information compromise is typically not feasible.

In contrast, the actual *probability* of developing lung cancer can be calculated as evidenced by the above figure. The cumulative probabilities were based on data derived from large population studies, where the effect of smoking could be isolated from other phenomena that might affect the study results.

Finally, malicious information compromises are frequently the handiwork of a third party. There is simply no way to predict where, when, and/or why a given individual or organization will strike against a particular entity. Often attacks are simply opportunistic, and a motive other than a chance to create havoc is not immediately obvious.

Notwithstanding these limitations, historical data on cyberattacks are not useless. Such attacks can be parsed to yield trends and a more informed view of the threat landscape. These data can expose the popularity of various attacks, which can help in crafting a defensible cybersecurity risk management strategy.

For example, the annual Verizon report is a genuinely useful if limited resource.[10] Although the data contained therein could be used to formulate a probability distribution of attack types, the predictive value of such a distribution is limited principally because of the inherent variability of IT environments.

In other words, simply because 30% of historical compromises have been denial-of-service attacks does not imply there is a 30% probability any given IT environment will experience the same fate. The Verizon study and similar studies can only provide general guidance on the likelihood of a future security incident. IT environments resist generalization with respect to predicting the probability of a particular attack.

In stark contrast, the curves in the above figure prescribe the odds of *anyone* developing lung cancer as a function of the duration of their smoking habit. A representative sample of smokers can be used to generalize to individual smokers in the general population.

This book is certainly not the first publication to suggest complexity is relevant to cybersecurity risk. For example, traditional complexity theory, which is not discussed

[10] Op. cit. 2020 Verizon Data Breach Investigations Report (DBIR).

in this text, has been applied to cybersecurity. According to at least one paper, lessons derived from the Animal Kingdom suggest diversity is the key to cyber-resilience.[11]

Although the identification of complexity as a driver of cybersecurity risk is spot on, diversity as a remedy applies specifically to thwarting external attackers. An attacker should be presented with as complex an environment as possible in order to decrease the probability of correctly identifying critical information assets. In other words, external attacks are frequently exercises in guessing, and therefore guessing should be made as difficult as possible.

In contrast, internal processes and configurations should be made as uniform as possible with the objective of creating more predictable environments. The intent is to decrease the likelihood of incomplete, incoherent, and/or inconsistent security risk management. The challenge from an internal perspective is to reduce the potential for information compromise by reducing the number of deviations from controlled conditions.

The analogy with the biological world also has inherent limitations. Clearly, biological and cybersecurity systems are not managed the same way. Creationists' claims notwithstanding, an active hand by a Supreme Being seems unlikely absent scientific evidence to the contrary. Adaptation in the natural world is born of necessity and implemented via a powerful combination of time and randomness.

These two factors are critical features of the competition for survival in intensely unforgiving ecosystems. Speciation, the result of genetic and epigenetic changes, is part of Nature's risk management strategy.

IT environments require intervention by humans who must continuously apply security controls to a multiplicity of risk factors for information compromise. These risk factors relate to the electronic information being processed, the technologies that support processing, and the network users who actively engage with both. Clearly, the spectrum of cybersecurity risk factors for information compromise has an effect on the magnitude of cybersecurity risk.

The key point is that the effect of these risk factors in aggregate can only be appreciated when the entire IT environment is in view. Simple as it may seem, this revelation represents a significant departure from traditional approaches to cybersecurity risk management.

However, developing a macroscopic view of cybersecurity risk is non-trivial unless simplifications are possible. One such simplification is security risk management is assumed to be a random variable thereby opening the door to statistical analyses. Such an assumption might appear frivolous since security professionals clearly make thoughtful and reasoned decisions on risk management.

One might counter this objection by arguing the make-up of IT environments nearly guarantees security risk management is not entirely deterministic. In fact, it has been argued that elements of chance underlie all events in the universe.[12]

---

[11] G. Ghandi, Complexity Theory in Cybersecurity; https://www.researchgate.net/publication/263652176_Complexity_theory_in_Cyber_Security, 2014.

[12] M. Eigen and R. Winkler, *Laws of the Game; How the Principles of Nature Govern Chance*; Princeton University Press, 1981.

> Everything that happens in our world resembles a vast game in which nothing is determined in advance but the rules, and only the rules, are open to objective understanding. The game itself is not identical with either its rules or with the sequence of chance happenings that determine the course of play. It is neither the one nor the other because it is both at once. It has many aspects as we project onto it in the form of questions. We see this game as a natural phenomenon that, in its dichotomy of chance and necessity, underlies all events.

It is clear no cybersecurity professional would admit to tossing a coin to make security decisions if they had any semblance of pride and/or a desire to remain gainfully employed. Therefore, it is understandable why there might be resistance to accepting a coin toss as a model for cybersecurity risk management.

However, a model does not have to be exact or even correct all the time to offer operationally useful lessons. Furthermore, simplifications are inevitable if a model is to describe inherently "complex" scenarios noting we have yet to actually define this word. The limitations that accompany such simplifications must be acknowledged, and care must be exercised to not over-generalize the results. Equally, generalizations that facilitate even minor insights into cybersecurity risk management should be embraced and refined over time. Developing such insights is precisely the goal of this book.

Contents

About the Author

Carl S. Young is the author of four previous reference books on science applied to security. He has held senior security risk management positions in government, banking and academia, and is currently the president and co-founder of Consilience 360, a security risk consulting firm based in New York City. He received his undergraduate and graduate degrees in mathematics and physics from the Massachusetts Institute of Technology (MIT).

Part I
Security Risk Fundamentals

Chapter 1
Core Concepts

1.1 Introduction

Unsurprisingly, any discussion on security risk management is based on the concept of risk. The fact is risk is broadly misunderstood, which may partly explain historic difficulties in its assessment. The casual use of terminology might contribute to this misunderstanding, although it is unclear if a lack of linguistic rigor is actually a cause or an effect. Whatever its origins, evidence of confusion is commonplace. For example, conflation of basic terms like "threat" and "risk" occurs frequently, even among security professionals.

The risk assessment process itself is likely affected by the confusion over risk fundamentals. A misunderstanding of the basic terms portends badly for rigorous risk assessments and increases the odds of a poorly formulated risk management strategy.

Assessing cybersecurity risk is best accomplished by initially adopting the broadest perspective. Such a vantage will increase the likelihood the maximum number of scenarios will be addressed with the minimum number of assessment criteria. Details should be considered once the general categories or relevant "food groups" have been identified. In general this practice works when evaluating multi-faceted problems.

A fundamental aspect of security risk management is that all threat scenarios are equivalent when viewed from a sufficiently high level. This equivalence partly explains why the risk assessment *process* is universally applicable. However, equivalent is the not the same as identical. Clearly all threat scenarios are not identical, which explains why the expertise required to manage security risk depends on the scenario details.

For example, the skills required to treat cancer are different from those needed to protect an IT environment. Although oncologists and cybersecurity professionals are both risk managers, they each receive specialized training that enables them to address their respective threat scenarios.

C. S. Young, *Cybercomplexity*, Advanced Sciences and Technologies for Security Applications, https://doi.org/10.1007/978-3-031-06994-9_1

If all threat scenarios are equivalent when viewed at a sufficiently high level, it begs the question of what drives differences in the magnitude of security risk across scenarios. In particular, how does one cybersecurity threat scenario differ from any other? To answer this question requires a more detailed examination of IT environments and threat scenarios.

1.2 IT Environments Versus Threat Scenarios

An IT environment can generally be described as a network(s) consisting of computers, systems and devices that enable information sharing. A cybersecurity threat scenario is simply an IT environment with the potential for information compromises and/or information-related business disruption.

In the real world, which means any network connected to the internet or an equivalent information sharing mechanism, the potential for such incidents exists in every IT environment. The implication is IT environments and cybersecurity threat scenarios are functionally equivalent.

Note a threat does not actually need to occur for an IT environment to qualify as a threat scenario. There merely needs to be the potential for such an occurrence, which suggests there is a relationship between threats and the conditions that make certain threats more likely or successful. In order to understand this relationship it is necessary to define threats as well as to describe IT environments and threat scenarios in more detail.

A threat is defined as anything with the potential to cause harm, damage or loss. Unsurprisingly, implicit in any phrase that includes the word threat is the presumption of a negative outcome. If there is no potential for harm, damage or loss, a scenario is not *threatening* by definition. Threats are manifest as threat incidents, and any entity that has experienced a threat incident must be worse off after an incident has occurred, again by definition.

Opinions can and do vary on what constitutes a threat, where the view often depends on the perspective of the viewer. Individuals can possess very divergent views on what constitutes harm, damage or loss as evidenced by the current political climate in the United States and elsewhere.

The canonical threat scenario is comprised of two elements in addition to the threat itself. These elements include an entity affected by a threat, which predictably is labeled an "affected entity." In the absence of an entity affected by or even potentially affected by a threat, a threat is merely an abstraction.

Another threat scenario element is the environment where threats and affected entities intersect. Intersection implies the threat and affected entity overlap in both location/position ("space") and time. The adjective referring to both space and time is *spatiotemporal*. If a specific threat and a particular entity do not overlap spatially and temporally, that specific threat-entity pair does not constitute a *bona fide* threat scenario *within a given context*.

Of course, in another context that specific threat might indeed intersect with that particular entity, and therefore the pair's status as a legitimate threat scenario would be restored. It is clear a specific threat must have a relationship to a particular entity in order for it to be relevant in a given context. It turns out that relationship is the essence of risk.

Contextual details relating to threats, affected entities and the environments in which they intersect are what differentiate one threat scenario from another. However, we repeat for emphasis that the categories or feature types linking a specific threat to a particular entity are the same in any threat scenario.

To recap, a canonical threat scenario consists of three elements as follows:

1. A threat, i.e., anything with the potential to cause harm, damage or loss.
2. An entity affected or potentially affected by a specific threat.
3. The environment where a specific threat and a particular entity intersect.

We now briefly discuss each of these elements individually with respect to cybersecurity threat scenarios.

The spectrum of cybersecurity threats is well known. It includes but is not limited to hackers of varying sophistication, entities sponsored by nation-states and insiders. The latter are defined as entities with authorized access to information assets but who possess malicious intent. Each of the above "threat actors" has different motives and capabilities although the general methods they employ are similar.

A distinction should be made between cyber threats/threat actors and the "attack vectors" they use. These terms are often used glibly with the potential to obscure the fact that humans always constitute the threat in cybersecurity threat scenarios. Computers do not unilaterally attack humans nor each other as of this writing, the actions of HAL 9000 in *2001: A Space Odyssey* notwithstanding.

Techniques employed by humans and executed via machines are used to commit information compromises, business disruptions and/or financial theft. The techniques employed by humans are merely *modus operandi* and not threats *per se*. That said, it is sometimes necessary and/or informative to specify the attack method. The necessity is driven by the requirement for specific security controls, since a particular security control might only be effective against certain attack methods.

For example, the security controls relevant to a denial-of-service attack might differ from those used to defend against a phishing campaign even though the same attacker might be involved in each attack. A viable security strategy must account for security control differences that are driven by both the specific attack method and the contextual details associated with a particular threat scenario.

The third element of a cybersecurity threat scenario is the IT environment. Recall this is the element where threats and affected entities intersect. All IT environments consist of computers, applications, endpoints, mobile devices and networking infrastructure used to process and/or share electronic information. Although these elements are common to all IT environments, their configuration and sub-elements can vary considerably within and across environments.

Knowing the individual elements of a threat scenario as described above is necessary but not sufficient to address a particular threat. It is critical to understand the

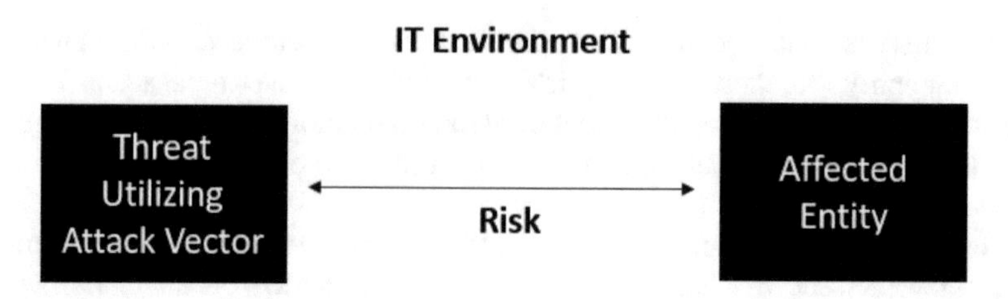

Fig. 1.1 A canonical cybersecurity threat scenario

relationship between threats and affected entities as well as the effect of environmental features on that relationship. As we hinted previously, this relationship is referred to as *risk*, and the scenario-specific *risk factors* that affect its magnitude are what drive the requirement for specific security controls.[1]

In view of this high level characterization, risk can be properly defined as follows,

> Risk describes the relationship between a specific threat and a particular entity within a given IT environment. Specifically, risk consists of three components: 1) the likelihood of a threat incident attempt or the potential for a successful threat incident, 2) the vulnerability or loss/damage if a threat incident did occur, and 3) the impact or loss-per-incident if a threat incident occurred. These three components determine the relevance of a specific threat to a particular entity within a given environment or context.

The components of risk are discussed in more detail in the next section. Figure 1.1 is an ultra-high-level representation of a canonical cybersecurity threat scenario.

1.3 The Components of Risk

Risk is not a tangible object, which may explain its quasi-abstract quality. Risk can only exist within a *bona fide* threat scenario, which implies all three threat scenario elements must be present to qualify for threat scenario bragging rights. In fact, the notion of risk only makes sense in the context of a legitimate threat scenario.

Unfortunately, the term "risk" has become synonymous with one of the three components of risk: likelihood. For example, a phrase such as "the risk of developing lung cancer" refers specifically to the likelihood of developing that disease. This parochial view can adversely affect the outcome of a security risk assessment if the other two components of risk are ignored and their magnitude is significant. That skewed outcome could in turn affect the strategy used to manage risk.

It is clear the relationship between threats and affected entities is influenced by the threat scenario details. Such details are what distinguish one scenario from another, and significantly, are what drive security risk management efforts.

Consider the likelihood of being attacked by a great white shark when swimming in the ocean versus standing on land. In the former environment the great white

[1] C. Young, Metrics and Measurements for Security Risk Management, Syngress, 2010.

shark is an apex predator capable of devouring nearly anything it fancies. The shark is *threatening* to humans because of its relative advantage in size, power, speed, and prodigious dental assets. But that advantage only exists in a specific context. Although this animal can devour nearly anything it encounters in the water, it is completely helpless if the same interaction occurs on land.

In other words, specific features of the threat scenario affect the *relevance* of the shark as a threat to humans. The potential for a shark attack is infinitesimally low on land because sharks can only survive in salt water.[2] This explains the insouciance at the prospect of being attacked by the real-life equivalent of Jaws while relaxing on the living room sofa.

It is therefore safe to say the threat of an attack by a great white shark is irrelevant to anyone comfortably ensconced inside their home. However, the threat becomes relevant after venturing into the great white shark's domain.

Although the interactions of sharks with humans is different than (say) those occurring between automobiles and pedestrians, the two threat scenarios are actually identical with respect to the types of elements present in each scenario. The same statement applies to any threat-entity pair.

Threat scenarios are even more nuanced since the relationship, i.e., risk, between threat scenario elements consists of three components. The three components of risk plus the three threat scenario elements must be present in order to qualify as a *bona fide* threat scenario. If any component or element is absent, a scenario is not a threat scenario by definition. Each component of risk will be discussed individually but the majority of the discussion will be devoted to likelihood because of its prominence in security risk assessments and its multiple incarnations.

We previously remarked on the colloquial use of the term "risk." In English the usage spans the linguistic spectrum, e.g., "I do not want to *risk* it." "What is the *risk* of contracting a virus?" "The financial transaction is too *risky*," etc. This linguistic flexibility has the effect of blurring the distinction between the three risk components. In all candor, the consequences might be meaningless to anyone except a linguistic scholar and a security professional. The latter's interest stems from the fact that ignoring one or more components of risk could affect the security strategy since each component might require a separate form of risk mitigation.

As noted above, one of the three components of risk is *likelihood*. Although it is not the only component, we maintain it is special because it has both a qualitative and quantitative interpretation. The distinction between these interpretations is significant, and the methods that apply in each case are not interchangeable.

The qualitative expression for the likelihood component of risk is more accurately referred to as the *potential* for a security incident. The potential refers to either a predisposition or propensity for a security incident to happen *or* the likelihood an attack will be successful *if* attempted. In other words, either conditions are ripe for an attempt or conducive to a successful attempt, where the interpretation will depend on the specific context.

---

[2] Some sharks can live in brackish water, e.g., bull sharks.

It is important to distinguish between these interpretations. Assessing whether an attack will be successful depends on issues such as any gaps in security controls. In contrast, assessing whether an attack will even be attempted is typically impossible to predict, especially with precision. The curves specifying the cumulative likelihood of developing lung cancer because of smoking actually specify the probability, i.e., the quantitative form of likelihood, a threat incident will actually take place. It is imperative to be precise about which version of likelihood is being evaluated since the analyses and resulting outcomes could vary significantly.

As we just noted, the *only* quantitative measurement of likelihood is the *probability* of a threat incident. This term is quite specific in terms of its meaning and the criteria for applicability. Importantly, calculating a probability requires a *probability distribution* of like items or events. Probability distributions are fundamental to the physical, chemical, biological, medical and social sciences, and therefore warrant further discussion.

A probability distribution is simply a fractional representation of a sample population. Probabilities are normalized to one by definition since each probability is a fraction of the total population. Of course, the sum of all the fractions must be unity. A probability greater than one or less than zero is meaningless.

A probability distribution is not a crystal ball. It merely enables generalizations based on a comparison of a sample of *similar* entities. Such entities might share a common characteristic such as in the previous example where a sample of individuals smoked tobacco. The comparison is possible because humans are anatomically and physiologically similar, which enables generalizations about other individuals who are not included in the sample but exhibit the same behavior.

Therefore, prerequisites for estimating the probability of a particular type of threat incident is a history of similar incidents and an assumption that conditions in the future will resemble the past. These prerequisites raise several questions that are fundamental to the theory of security risk assessments.[3] Specifically, how will the likelihood of future incidents be determined if there are no historical threat incidents and does the absence of historical threat incidents imply an absence of risk?

The relevance of historical threat incident data depends on whether previous incidents can be compared to future versions of same. If conditions relevant to the original data have significantly changed, any generalization based on historical data would be suspect.

Suffice it to say the inability to formulate a probability distribution of historical threat incidents has significant implications to assessing cybersecurity risk. Its relevance compels us to explore probability distributions in more depth, and to seek alternative methods in assessing the likelihood component of risk.

To illustrate the concept of a probability distribution, suppose a sample population of boxes consists of 100 boxes, which are identical except for their color. Twenty-five boxes are blue, 25 are red, 30 are green and 20 are yellow. Their respective frequencies, i.e., the fraction/percentage of the sample total, are indicated in Table 1.1.

[3] C. Young, *Risk and the Theory of Security Risk Assessment*, Springer Nature, 2019.

Table 1.1 Probability distribution of colors

	Fraction of the total population	Percentage of the total population	Probability of random selection
Blue boxes	25/100	25	0.25
Red boxes	25/100	25	0.25
Green boxes	30/100	30	0.30
Yellow boxes	20/100	20	0.20
Total	100/100	100	1.0

If selecting a box were a random variable, the probability of selecting a particular box color from among peer boxes would equal the fraction each color occupies within the sample population. In fact, the percentages of box colors specified in Table 1.1 constitute a perfectly respectable probability distribution, which is graphically illustrated in Fig. 1.2.

The population of boxes in Fig. 1.2 has been lumped together to facilitate comparison based on box color. If the population consisted of boxes, cats, dogs and sport shirts, the latter three categories must be excluded from the distribution since only the association between boxes and colors is relevant to the analysis. For example, if the probability of identifying a yellow box from the sample were the objective, a sample consisting of sports cars, sport shirts and farm animals would be irrelevant irrespective of their color.

Every probability distribution is characterized by its mean, i.e., average value, and its standard deviation, which is the spread or dispersion about the mean. The standard deviation specifies the average distance of the sample population from the mean value. Equivalently, it depicts the uncertainty in the mean value of the distribution. The variance of a distribution equals the standard deviation squared.

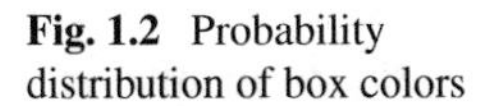

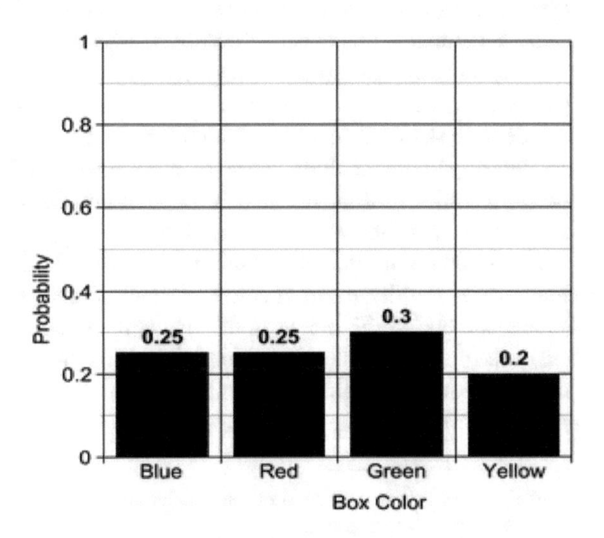

Fig. 1.2 Probability distribution of box colors

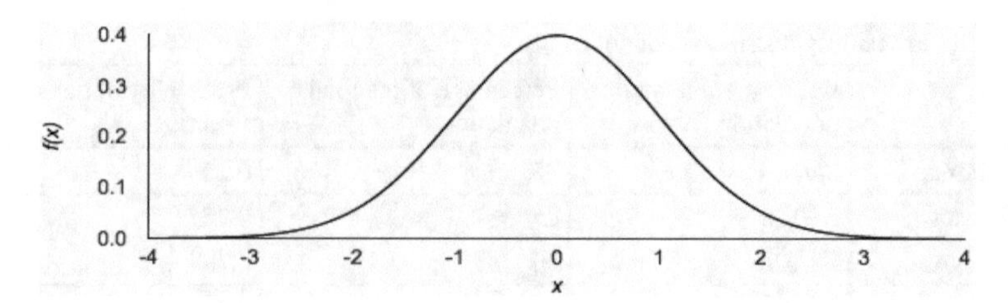

Fig. 1.3 The standard normal distribution

For example, the mean of the distribution depicted in Fig. 1.2 is 0.25, and the standard deviation is approximately 0.04. The smaller the standard deviation the tighter the distribution is clustered about the mean value. In other words, the spread of values about the mean is larger for distributions with larger standard deviations. However, the probabilities or fraction of the total distribution corresponding to one, two, three, etc., standard deviations are the same for identical types of probability distributions, e.g., normal, Poisson, etc.

One key objective of formulating a probability distribution of a sample population is to understand the make-up of the population from which a sample is drawn, i.e., the parent population. One might argue it is the motivation for statistics in general.

For normally distributed random variables, the standard deviation of the parent population equals the standard deviation of the sample divided by the square root of the sample size. Larger sample sizes result in greater statistical confidence in the uncertainty about the mean of the parent distribution. For example, the standard deviation of a normally distributed population of 200 elements will be $\sqrt{(200/100)} = 1.41$ times less than a sample of 100 elements drawn from the parent.

Most people have at least a passing familiarity with the normal distribution or "bell curve," which is named for its classic bell shape. Normal distributions are also referred to as Gaussian distributions, named after the nineteenth century German mathematician and physicist, Carl Friedrich Gauss.

As noted above, the spread or dispersion of identical types of probability distributions is always the same. In other words, if the fraction of a sample normal population includes one standard deviation from the mean, i.e., to the right and left of the mean, this sub-population includes 68% of the entire sample. Two standard deviations always includes 95% of the sample population. Three standard deviations always includes 99.1% of the sample population.

The standard normal distribution is a normal distribution that has been re-parameterized with a mean of zero and standard deviation of one. Tables specifying the precise relationship between the standard deviation and the corresponding fraction of the population have been published. This distribution is useful for estimating statistical confidence and other statistical metrics, e.g., the *p*-value. The standard normal distribution is depicted in Fig. 1.3.[4]

[4] This figure was drawn using an on-line applet available at https://homepage.divms.uiowa.edu/~mbognar/applets/normal.html.

It turns out all probability distributions tend toward normality as the sample size increases. A remarkable mathematical result is that if repeated samples are drawn from a parent distribution it will yield a normal distribution even if the parent itself is not normally distributed. This result is known as The Central Limit Theorem.

In a normal distribution of sample size N the standard deviation equals $\sqrt{N}$. From this fact it is possible to determine the sample size required to ensure a specific dispersion about the mean, i.e., the selection precision. It turns out precision scales as N^2 as we will demonstrate next.

Assume the required precision in selecting a distribution element equals one tenth the sample size, i.e., $\sigma = N/10$. Since $\sigma = \sqrt{N} = N/10$, solving for σ yields $N = 100$. Therefore, a sample size of 100 is required to achieve a precision within one tenth the sample size.

What if the required precision is N/100 or one hundredth the sample size? Using the same reasoning, the sample size N must equal 100^2 or 10,000. Therefore, increasing the precision requirement by a factor of 10 requires squaring the sample size. The N-squared precision dependence can have significant implications to statistical sampling.

Much ink has been devoted to the likelihood component of risk, but we dare not ignore the second and third components if estimating the magnitude of risk in totality is required. Recall these components of risk are *vulnerability* and *impact*.

Vulnerability is the potential for loss or damage if a threat incident occurs. The impact component of risk is the significance of the threat incident to the affected entity or some other interested entity. Vulnerability and impact are related, where the impact can be defined as the vulnerability-per-threat incident.

The following scenario illustrates the distinction between the vulnerability and impact components of risk. Assume an individual is carrying $10,000 in cash. If that individual is the affected entity in a robbery threat scenario, the magnitude of the vulnerability component of risk is precisely $10,000. This amount reflects the total amount of money that could be lost if a successful threat incident occurred. If that affected entity's entire life savings equaled $20,000, the impact component of risk is significant since the loss constitutes half the victim's net worth.

Now consider a similar but slightly amended threat scenario with significant implications. In the new version the affected entity is a multi-billionaire who also happens to be carrying $10,000 in cash. In this scenario the vulnerability component of risk is exactly the same as before, but the impact is negligible since the billionaire would presumably hardly notice a loss of $10,000.

Note the threat scenario concerns the threat of robbery. If the threat scenario is focused on physical violence, the vulnerability and impact components of risk would presumably be similar if not identical for both individuals. The difference in the two scenarios is clearly significant, and the difference highlights the importance of articulating the specific threat in addition to the particular component of risk being assessed.

As noted previously, all three components of risk must be present in a scenario order to qualify as a *bona fide* threat scenario. The magnitude of risk is expressed as the product of the three components and is sometimes called the Fundamental

Expression of Risk,

$$\text{Risk(threat scenario)} = \text{Likelihood} \times \text{Vulnerability} \times \text{Impact} \qquad (1.1)$$

The general form of (1.1) merely expresses the fact that the risk associated with a particular threat scenario always consists of three components. However, the form of this equation is purely indicative. For example, (1.1) is linear in all three components of risk yet every threat scenario is not necessarily linear with respect to the magnitude of the likelihood, vulnerability and impact components of risk.

In addition, a more precise expression of risk might include a term in the denominator reflecting the moderating effect of security controls. All threat scenarios contain security controls. Even the mere awareness of a threat might qualify as a form of control.

Strictly speaking, security controls have no effect on the risk inherent to a given threat scenario. To reiterate, risk describes the relationship between a specific threat and a particular affected entity within a given environment. However, security controls must be included in a representation of *residual risk*, which is a concept discussed later in this chapter. It is often difficult and/or impractical to distinguish between a threat scenario with and without security controls. Therefore, residual risk and inherent risk are operationally equivalent.

The upshot is although (1.1) specifies all components of risk for any threat scenario it does not constitute a proper equation nor does it specify the precise magnitude of risk. Naively multiplying the values of the three components and arriving at a single figure actually makes no sense. Each component represents a distinct quantity whose measurements are mutually incompatible assuming such measurements were even possible.

Consider the special case where the likelihood component of risk can be expresses as a probability, which is by definition equal to a numerical value between zero and one. The vulnerability component of risk specifies the magnitude of loss, which could be expressed as revenue, the number of casualties, etc. The impact component of risk is often similarly expressed. Therefore, combining the disparate values into a single number that reflects the overall magnitude of risk is neither possible nor meaningful.

Importantly, security controls designed to reduce vulnerability or loss often do nothing to affect the likelihood of a threat incident. That distinction might be critical when developing a security risk management strategy. For example, a threat incident causing minimal financial loss could still affect the impact component of risk through reputational fallout. The lesson is to be specific about the precise threat *and* the component of risk being addressed.

1.4 Risk Factors and Risk-Relevance

To summarize, we now know that risk characterizes the relationship between a specific threat and a particular entity within a given environment. Moreover, a threat scenario always consists of the same three types of elements and risk always consists of the same three types of components. The magnitude of each component is determined by specific behaviors, conditions and/or features inherent to a given threat scenario, and are known as *risk factors*. Security risk management is defined as the application of security controls to risk factors.

Notably, the risk factors can vary by component for a given threat scenario. In other words, the risk factors associated with the likelihood component of risk often differ from those affecting vulnerability or impact. For example, assets stored within a home is a risk factor for the vulnerability component of risk with respect to the threat of burglary. The quality of the deadbolt on the front door is a risk factor for the likelihood of a *successful* burglary. A sturdy deadbolt in combination with a solid door and door frame might also dissuade a burglary *attempt*. The time of day and home location are also risk factors for the likelihood of a burglary attempt.

It is necessary to determine the specific risk factors for each component of risk because the security controls required to address each risk factor can vary accordingly. Furthermore, the need for different security controls might exist irrespective of whether the risk factors relate to the same or different components of risk.

The concept of a risk factor leads to the notion of *risk-relevance*, noting the word relevance has already surfaced several times. Simply put, if a given threat scenario feature increases the magnitude of risk, which is equivalent to saying a feature is a risk factor for one or more components of risk, it is deemed risk-relevant. Risk-relevance is a useful term because it implicitly relates threat scenario features to components of risk. In other words, it succinctly expresses whether a particular threat scenario feature contributes to the magnitude of security risk.

1.5 Residual Risk

Security controls can never reduce the magnitude of risk to zero. There is always residual risk in any threat scenario, where residual risk is defined as the magnitude of a component of risk following the application of security controls.

The persistence of residual risk motivates use of terms like "risk reduction" and "risk management" rather than "risk elimination." Risk management doesn't eliminate risk, but rather it is intended to reduce the magnitude of one or more components of risk to some acceptable level. That level is dictated by the inherent limitations of security controls and the organizational tolerance for security risk. The latter concept is critical to strategic security risk management and is discussed in detail in Chap. 9.

Resource constraints impose practical limits on the scope of security risk management efforts. Inevitably some risk factors will only be partially addressed or might be completely ignored depending on what other risk-relevant issues are present and available resources.

The security risk management process is recursive, where security control adjustments are regularly performed in response to changes to a dynamic risk profile. In other words, risk management consists of periodic risk assessments that inform

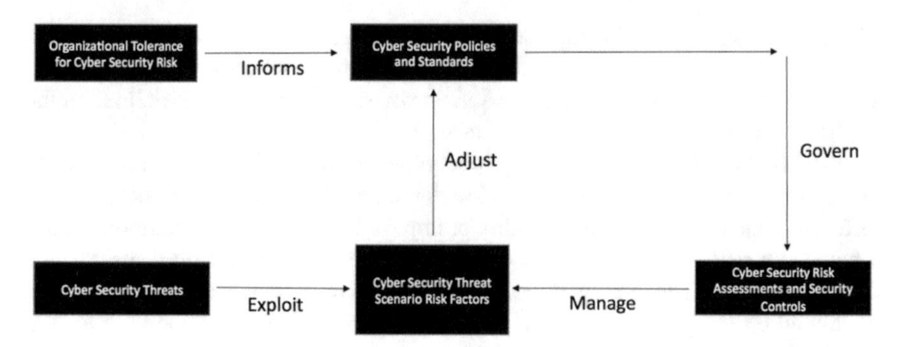

Fig. 1.4 The cybersecurity risk management process

regular adjustments to the magnitude of residual risk. Figure 1.4 illustrates the feedback loop that characterizes the cybersecurity risk management process.

Note that cybersecurity policies and standards are explicitly specified in the feedback loop. These security controls in particular are fundamental to rigorous security risk management since they dictate the acceptable limits on user behavior and technology performance. Security technology standards and cybersecurity policies ultimately reflect the organizational tolerance for risk, a concept that has already surfaced since it is fundamental to security risk management. Policies and standards also promote uniformity, which we will see is an antidote to complexity in IT environments.

1.6 Risk Assessment Universality

The fact that risk always consists of the same types of components and all threat scenarios consist of the same types of element is significant. The significance relates to the fact that *the risk assessment process is identical for any threat scenario.* In other words, there is a standard approach to security risk management, which is represented by the feedback loop depicted in Fig. 1.4.

Although the risk assessment process has universal applicability this does not imply all threat scenarios are identical. Clearly the likelihood of experiencing a blizzard is different in Alaska than the Gobi desert. It means the *approach* to security risk management is always the same irrespective of the scenario details.

However, the details are relevant to determining who should perform the assessment and how the threat scenario risk factors should be addressed. For example, a computer security expert would not be the obvious choice to assess the risk of hurricanes in Florida. Similarly, meteorologists would not be at the top of the list to evaluate the potential for information compromises within IT environments.

However, both professionals utilize (or should do) the same approach in assessing the risk associated with the respective threat scenarios of concern. Namely, they each examine and prioritize the risk factors for the three components of risk associated with their respective threats of concern. Risk management follows, where mitigation is applied according to the prioritized results and in consideration of the financial and operational costs of mitigation.

1.7 Risk Calibration and Variability[5]

Physical systems can present useful analogies for security-related scenarios, although care must be exercised not to push the limits too far. For example, the use of statistics to describe physical systems with many atoms or molecules inspired the following discussion.

Scientists have identified a thermodynamic property called temperature, which characterizes the average kinetic energy of a macroscopic system of particles. Such a system can be described in terms of microstates, where a particular configuration of particles corresponds to a specific microstate.

Thermal contact between two systems can be described statistically, where the absolute temperature (more precisely, the reciprocal of the absolute temperature) reflects the relative change in the number of microstates of a macroscopic system as the energy changes and assuming the volume remains constant.

The concept of microstates can be loosely applied to IT environments that contain numerous risk factors. The risk factors are analogous to atoms/molecules in the physical world. The analogy specifically relates to the number of atoms/molecules and risk factors in their respective environments and the difficulty associated with analyzing individual elements within each environment.

Both environments contain a huge number of elements, and the intractability of assessing each element is what necessitates a probabilistic approach. Moreover, such an approach is key to facilitating a macroscopic view of each environment. However, differences between physical scenarios and security threat scenarios also highlight the difficulties inherent to assessing security risk on an enterprise, i.e., macroscopic, scale.

It is significant that only a single measurement of temperature is needed to determine the average kinetic energy of an isolated object in equilibrium with its surroundings. As the reader is undoubtedly aware, a thermometer is an instrument that measures temperature, and thermometers are calibrated with respect to well-known scales such as Fahrenheit and Celsius. These scales are based on the freezing and boiling points of water. Although a particular temperature reading is typically an unheralded event, such readings would be useless if the thermometer were not calibrated with respect to a recognizable and interpretable scale.

What is meant by calibration? Substances formerly used in thermometers such as mercury react in a repeatable way under similar thermodynamic conditions, which

[5] The content on temperature, thermometers and density of states paraphrases the discussion in *Fundamentals of Statistical and Thermal Physics*, C. Reif, McGraw Hill, 1965.

results in the same reading relative to the scale of choice. Recognize that the chosen scale is arbitrary but measurement readings must be consistent for a given temperature. The consistent behavior of a thermometer under identical thermodynamic conditions is fundamental to temperature measurements.

For example, consider a Thanksgiving turkey, Christmas ham, and Passover brisket cooking in an oven set to 400 °F. Each item should measure the same temperature if the same thermometer is used and that thermometer is accurately calibrated relative to the Fahrenheit temperature scale. A thermometer would be of little use if temperature readings were affected by the material being measured or the temperature readings were inconsistent.

Physical systems are generally comprised of a huge number of molecules. Even a teaspoon of liquid contains on the order of 10^{23} molecules. A thermometer will actually come in contact with only a small fraction of the molecules in a system or sub-system yet a temperature measurement is indicative of the system temperature as a whole.

Stated more precisely, the entire system has an energy E, which could be measured as small intervals or energy levels, ΔE. As indicated above, a statistical characterization of the system is facilitated by describing it in terms of microstates, where each energy level will contain many possible microstates within the overall system, even for physically small systems.

The number of *accessible* microstates of a physical system is particularly relevant if a statistical view of matter is adopted. Suppose a particular system is comprised of many atoms or molecules. Each configuration of the atoms or molecules in the system can be described as a microstate. Accessible microstates result from constraints imposed on the system, e.g., the total energy.

For example, if the atoms or molecules are constrained to occupy a fraction of the total volume of a given environment it limits the number of accessible microstates. If the system is isolated and in equilibrium, each accessible microstate is equally likely. In other words, under such conditions the number of probable states is a maximum.

Figure 1.5 depicts a physical system with eight particles initially constrained to inhabit volume V_1. Keeping the kinetic energy constant but doubling the volume to $V_2 = 2V_1$ will increase the number of accessible microstates by a factor of 2^8 since the number of available volume elements per particle has doubled for all eight particles.

More generally, the total number of accessible microstates, Ω in an isolated physical system is related to a thermodynamic state function known as *thermodynamic entropy*. The thermodynamic entropy (S) is directly proportional to the natural logarithm of the number of microstates, Ω. Although conceptually similar, thermodynamic entropy should not be confused with information entropy, where the latter concept is fundamental to information theory and the model of complexity described in later chapters.

The probability of being in any particular physical state equals the ratio of the number of microstates to the total number of accessible microstates. Changing a system constraint, e.g., the total energy, so that the system is no longer in equilibrium

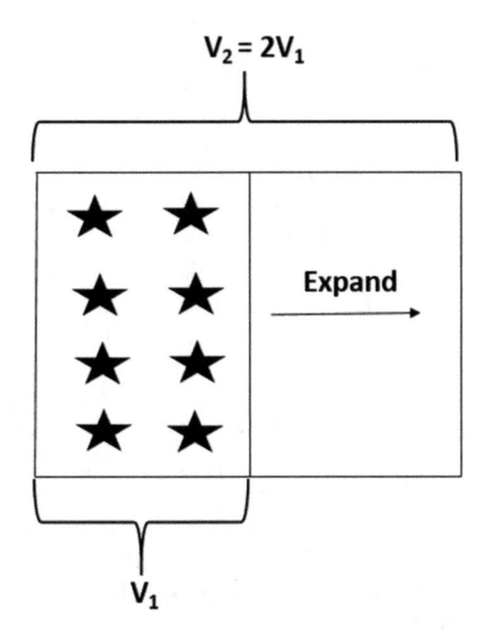

Fig. 1.5 A change in the number of accessible microstates by increasing the volume

with its surroundings will reduce this ratio so that a particular microstate is more probable than its peer states.

A fundamental law of thermodynamics states that the thermodynamic entropy of isolated systems always increases. In other words, the total number of accessible microstates of isolated systems always increases. To reiterate, the number of accessible microstates of a system in equilibrium is a maximum, which defines an equilibrium condition.

Energy must be added or subtracted from the system to create a condition where the probability of a particular accessible microstate is higher than all others. If left to its own devices and allowed to equilibrate without absorbing energy, the system will eventually achieve a condition where all accessible states are equally likely, i.e., equilibrium.

An equilibrium condition also relates to system temperature. If the thermometer and the system under measurement are in equilibrium, they have the same temperature by definition. At this temperature, the probability the system and thermometer are in a particular microstate with the identical energy, relative to all other energies, is a maximum.

A statistical description of physical systems reveals why an actual measurement of security risk similar to temperature cannot be made. Physical systems in equilibrium with each other will exist at the same temperature according to the Zeroth Law of Thermodynamics. It does not matter where the measurement is made since the system is isolated and in equilibrium.

In contrast, an IT environment exhibits significant variability over small physical, virtual and/or temporal measurement scales. Therefore, a hypothetical measurement of its status in one location will not necessarily correlate with such a measurement made elsewhere in the environment.

Moreover, the notion of equilibrium in IT environments is of no practical consequence because spatiotemporal time and distance scales are short compared to corresponding security risk assessment scales. A statistical treatment of risk factors appears to be the only realistic option for assessing risk on an enterprise scale, i.e., across the entire IT environment.

The analogy with physical systems reveals another obstacle to assessing cybersecurity risk: the inability to calibrate security controls. The variability of IT environments coupled with the lack of a concrete reference precludes fine tuning the magnitude of security risk by adjusting one or more security controls.

Finally, we note the principal benefit of applying statistical methods to assessing cybersecurity risk in IT environments is also its most significant drawback. Namely, such methods cannot assess the effects of individual risk factors on the potential for information compromise. The ineluctable conclusion is *both* microscopic and macroscopic assessments are necessary to determine the magnitude of cybersecurity risk in totality. The contention is it is crucial to appreciate the benefits and limitations of statistical methods applied to IT environments while leveraging any insights so derived.

Chapter 2
Representing Cybersecurity Risk

2.1 Introduction

A picture is said to be worth a thousand words. The same can be said of a trend line. More precisely if less pithily, a trend line reveals how a function is changing in both magnitude and direction. In mathematical terms, a trend line graphically reveals the relationship between the independent and dependent variables of a function. If the independent variable happens to be a specific threat scenario parameter or feature, e.g., a risk factor for information compromise, and the dependent variable is a component of risk, risk-relevant information is conveyed at-a-glance.

There are multiple ways to display trend lines and quantitative information in general. The particular method depends on the data, the message to be conveyed and the intended audience. Often trend lines show the change in some variable as a function of distance or time. One particularly memorable depiction of change was the use of line width to plot the ever-dwindling size of Napoleon's forces during the Russian campaign of 1812.[1]

Unfortunately, the identification of risk-relevant trends is a rare outcome of a cybersecurity risk assessment. This situation is not an indicator of the importance of trends. It is actually cause for lament since a trend line between risk-relevant variables can inform a risk management strategy.

It is sometimes possible to infer such a relationship. The result is a directionally meaningful if less precise understanding of the magnitude of a component of risk. Whether quantitative or qualitative, understanding the relationship between risk-relevant variables within IT environments can provide insights into the likelihood, vulnerability and/or impact of information compromises.

As expressed in the final paragraph of Chap. 1, both microscopic and macroscopic views of IT environments are necessary to develop a comprehensive understanding of cybersecurity risk. The required view will be affected by the scale used to perform the assessment and the choice of scale is tied to the vulnerability being assessed.

[1] Edward R. Tufte, *The Visual Display of Quantitative Information*, Graphics Press.

C. S. Young, *Cybercomplexity*, Advanced Sciences and Technologies for Security Applications, https://doi.org/10.1007/978-3-031-06994-9_2

This is a fundamental result of the theory developed in this book and its applicability transcends cybersecurity risk management.

For example, resolving objects on the moon from earth requires a telescope with a lens of sufficient aperture, where the aperture size will depend on the size of the objects to be resolved. In contrast, a microscope is needed to view cellular structures that are much closer but might be one millionth the size of the objects located on the moon. Neither instrument is much use in resolving objects for which they were not designed. Optical resolution is governed by the diffraction limit, and the physical dimensions of objects observable via optical instruments such as the human eye, telescopes, etc., are tethered to this limit.

The required level of detail also dictates the scale or perspective needed to observe non-physical phenomena. Extending the analogy to the very limit of reasonableness, the "lens" used to resolve a single line of code within an application will be different than the one used to analyze an IT environment consisting of thousands of devices and users. It is difficult to imagine how an assessment of the potential for information compromise would yield the same result in each case.

Physical lenses are clearly not directly relevant to IT environments. But the point is the scale used to observe non-physical phenomena such as those occurring within IT environments will also affect assessment results. Specifically, the choice of scale in identifying and displaying risk-relevant phenomena is risk-relevant.

Finally, the question is not whether the magnitude of cybersecurity risk varies with the assessment perspective. It seems obvious that assessment results will, at a minimum, depend on the number of risk factors surfaced during an assessment. The key questions are precisely *how* the magnitude of cybersecurity risk is affected on an enterprise scale and what are the specific IT environment dependencies. To fully address these questions requires an understanding of the behavior of functions that describe risk-relevant relationships.

2.2 Linearity and Non-linearity

The previous discussion suggests the need to be more precise about the nature of relationships between risk-relevant parameters. A prerequisite to appreciating the implications of risk-relevant relationships within IT environments is to understand the distinction between linearity and non-linearity.

Assume a mathematical function, f(x, y), consists of two variables, x and y. In this function x is the independent variable, and therefore it can assume any value. The variable y is known as the dependent variable, and its value *depends* on the value of x. A *linear* relationship between x and y implies a change in the value of x causes a *proportionate* change in y.

In other words, if the value of x is doubled the value of y is doubled, if the value of x is tripled the value of y is tripled, etc. As its name implies, a graphical representation of a linear relationship between variables will appear as a straight line.

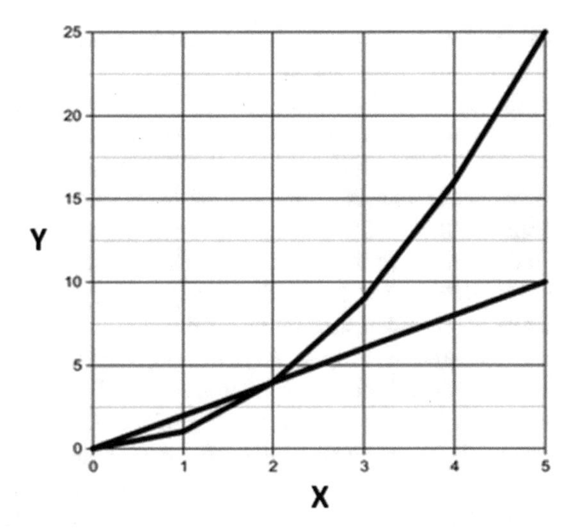

Fig. 2.1 Linear and non-linear relationships between variables

In contrast, a *non-linear* relationship between two variables x and y in a function f(x, y) means a change in the value of x results in a *disproportionate* change in the value of y. Therefore, as the value of x increases the value of y does not merely grow in kind. Rather, the disparity between successive values of y accelerates with increasing values of the independent variable, x.

The two curves in Fig. 2.1 depict a linear and non-linear relationship between the variables, x and y.

The straight line in Fig. 2.1 is a plot of the expression $y = 2x$. The curved line is a plot of y equals x times x or equivalently, $y = x^2$. The integer two in the latter expression is the exponent of x. An exponent specifies how many times a variable is multiplied by itself. The exponent associated with the independent variable in a function is typically the critical parameter in determining the rate at which a function is changing.

For example, x-squared equals x times x or x^2, x-cubed equals x times x times x or x^3, etc. The magnitude of the exponent determines the degree of non-linearity and hence the steepness of the curve. Therefore, the exponent of x determines the rate of growth (or decay if the exponent is negative) of y.

The integer two in the linear expression is a coefficient of x and therefore simply multiplies the values of x. Note the exponent is implicitly equal to one in the linear expression. In other words, the expression $y = 2x$ is equivalent to $y = 2x^1$. A condition of linearity implies the exponent of the independent variable equals one.

The relationship between risk-relevant variables can have deleterious implications depending on the scenario. Assume the exposure to a radioactive isotope is a concern. Let "t" equal exposure-time and "n" equal the number of cancer deaths as a function of exposure-time. If the relationship between n and t is linear, one could determine the acceptable limits of exposure simply by extrapolating from a few initial data points. However, if this relationship is non-linear, restricting the analysis to a short initial time interval would not reveal the full biological effect of prolonged exposure with potentially tragic results.

A linear versus non-linear relationship between risk-relevant variables also has potentially significant implications to IT environments. An obvious example is the effect of the number of risk factors for information compromise on one or more components of cybersecurity risk. If the particular relationship is linear, simply multiplying the results gleaned from one segment by a proportionate increase in the total number of segments will yield an accurate result for the entire IT environment. However, if the relationship is a non-linear function such a multiplication could severely underestimate the result.

It is crucial to identify such relationships in IT environments, and it is especially important to determine if such relationships are linear or non-linear. Even a qualitative appreciation for the nature of the dependence is helpful in developing an effective security risk management strategy. Estimating the broader effects of a condition or phenomenon based on a sample or subset is called extrapolation. The validity of an extrapolation depends strongly on whether the effect being extrapolated behaves linearly or non-linearly and the extent of the extrapolation.

Finally, the scale used to assess IT environments will affect the outcomes of cybersecurity risk assessments, and is particularly impactful to environments with non-linear dependencies on risk-relevant variables. The scale-dependence of cybersecurity risk is a key result arising from the stochastic model of cybersecurity risk management discussed in Chaps. 6, 7 and 8.

2.3 Security Risk Models

A cybersecurity risk assessment cannot possibly account for every IT environment risk factor if the exclusive focus is on individual vulnerabilities. The only recourse is to develop a *model* of cybersecurity risk that can approximate the aggregate effect of risk factors. The model would ideally specify a quantitative relationship between risk-relevant variables that accounts for the aggregate effect.

If the relationship is linear, a simple multiplication will yield accurate estimates of the dependent variable across relevant values of the independent variable. If the relationship is non-linear, a simple multiplication will cause the dependent variable to become increasingly inaccurate with increases in the independent variable. Therefore, it is particularly important to identify non-linear effects in cybersecurity risk assessments.

In practice, relationships between risk-relevant variables can be difficult to identify and even more difficult to quantify. However, it is sometimes possible to estimate such relationships and thereby infer how changes in a risk-relevant parameter will affect the potential for information compromise. Notwithstanding the inherent imprecision, such estimates might at least indicate the direction of risk-relevant trends, which can support decisions on security risk management.

In summary, identifying a quantitative relationship between risk-relevant parameters is of significant value in developing an accurate assessment of cybersecurity risk. However, even qualitative estimates of such relationships can help inform a cybersecurity risk management strategy. Non-linear relationships are particularly risk-relevant because of the potentially disproportionate effect on one or more components of risk.

2.4 Security Risk Categorization

The previous discussion revealed the importance of distinguishing between linear and non-linear phenomena in assessing cybersecurity risk. The behavior of such phenomena would ideally be characterized by a model defined over the range of values relevant to the scenario. What is the best way to identify such relationships when assessing cybersecurity risk within IT environments?

Since IT environments potentially contain an infinite number of risk-relevant features, there are infinite ways to characterize the potential for information compromise. One approach would be to specify all possible variations of risk factors and security controls and evaluate the resulting combinations and permutations. This approach is inefficient and would be ineffective for any realistic IT environment.

A more efficient representation of IT environments would be to identify the maximum number of risk-relevant scenarios using a minimum number of categories. This approach would characterize the environment from the most general vantage thereby maximizing the likelihood of encompassing the breadth of risk-relevant scenarios.

Subsequently focusing on the details could identify specific and/or idiosyncratic features that might require special security controls.

This approach would also help identify the security controls with the broadest applicability. Scenarios that fall outside the major categories could then be evaluated with respect to their frequency of occurrence and/or impact. In addition, resources can be optimized by prioritizing security controls based on the breadth and/or intensity of their effect.

In other words, a security risk assessment would be initially based on the largest categories or "food groups" of scenarios. Such an approach minimizes the likelihood of omitting a particular scenario. In addition, a broad perspective is more likely to connect the dots to expose a potential relationship between risk-relevant parameters. Of course, a significant challenge is to identify the relevant food groups.

Unfortunately the opposite tack is frequently taken when assessing cybersecurity risk. Namely, the initial emphasis is on identifying specific vulnerabilities. Such details can certainly be risk-relevant and should not be ignored without cause. However, it is prudent to begin with an expansive view followed by an increasingly narrow focus if a comprehensive view of cybersecurity risk is the objective.

The previous sections discussed the high-level approach to assessing cybersecurity risk and the necessity of identifying relationships between risk-relevant parameters. Quantitative relationships in particular can enable rigorous security risk management decisions. The precise behavior of these relationships and the implication to assessment perspectives will be explored next.

Chapter 3
Scale and Scaling Relations

3.1 Introduction

Readers old enough to recall using paper maps are already familiar with the concept of scale. A scale was printed on the map and it specified the ratio of map distance to actual distance. The bigger the ratio the coarser the scale and the less detail was visible. If the scale is too detailed the area covered by the map might be too limited. If the scale is too coarse, the requisite detail would be absent, which would potentially render the map useless for navigation.

For example, if one inch on the map corresponds to one mile of actual distance, the level of detail would be much less than if one inch of map distance corresponds to one inch of actual distance, i.e., a one-to-one scale. Zooming in and zooming is a feature of Google Maps that enables viewers to smoothly change the map scale over a range of values.

In Chap. 2 we remarked that the scale used to engage with the world affects the perception of the world. One must adjust the scale to match the particular scenario of interest, and that lesson applies to life in general. Focusing on details versus viewing the bigger picture is frequently a significant issue when assessing a situation or scenario.

Also recall from Chap. 2 that a trend line graphically displays the relationship between the independent and dependent variables of a function. The scale along each axis of a graph representing a function is actually equivalent to the scale of a map. In the case of a function the scale indicates the change represented by each incremental unit along the axis.

In other words, the scale of a function and the scale of a map are both indicative of change. For example, in the function $y = x^2$, we say y changes or *scales* as the square of x. The precise relationship between the x and y variables is known as a *scaling relation*, and the scale used to display the function is dictated by that relation. Identifying a proper scaling relation between risk-relevant parameters is the Holy Grail of security risk assessments because it can facilitate prescriptions for applying security controls.

C. S. Young, *Cybercomplexity*, Advanced Sciences and Technologies for Security Applications, https://doi.org/10.1007/978-3-031-06994-9_3

In exact analogy with a map, the scale of the axes of a graph depicting a trend line affects the detail being displayed. Therefore, the scale of the axes must be matched to the function in order to achieve the proper perspective. This requirement is a special case of the general requirement to engage the world from the proper perspective, and thereby capture the required detail.

3.2 Cybersecurity Risk and Perspective

Focusing on a single server or application is not likely to expose patterns affecting the entire IT environment. A broader perspective is necessary to reveal phenomena that are only manifest an enterprise scale. Furthermore, assessing a narrow segment of an IT environment will almost certainly result in an assessment of cybersecurity risk that differs from one gleaned from a broader perspective.

In addition, effects that are manifest on an enterprise scale would only be apparent if the scale of the assessment matches that view. Conversely, a broad perspective is not likely to identify individual risk factors for information compromise. As highlighted above, identifying software code vulnerabilities requires a detailed code review.

The perspective used to assess cybersecurity risk can be represented on a theoretical continuum. At one extreme is a microscopic view that reveals individual vulnerabilities. At the other extreme is a macroscopic view that accounts for the spectrum of risk factors and exposes phenomena manifest on an enterprise scale. Figure 3.1 depicts the so-called Cybersecurity Risk Assessment Continuum.

Each additional IT environment element will invariably include more risk factors for information compromise. A decrease in the potential for information compromise would be expected to accompany any risk assessment that excluded risk factors. Of course the excluded risk factors will not go away by simply ignoring them. In addition, specific risk factors might disproportionately affect the potential for information compromise if the relationship between those risk factors and the magnitude of the relevant component of risk were non-linear.

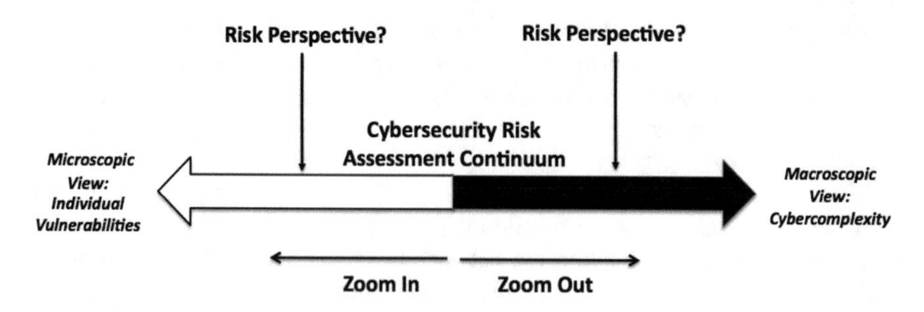

Fig. 3.1 The cybersecurity risk assessment continuum

Finally, the perspective used to assess an IT environment will be different depending on the vantage. An attacker is often constrained to hunt for targets of opportunity via random selection. Therefore, identifying a specific information asset becomes increasingly difficult as the number of assets increases.

The probability the attacker selects a particular asset at random scales inversely with the number of assets and therefore disproportionately decreases the potential for information compromise. In this case the uncertainty that accompanies a probability distribution of possible outcomes is an ally to security risk management efforts.

In contrast, the uncertainty associated with *internal* processes, workflows and technologies *increase* the potential for information compromise through behaviors and actions that deviate from the so-called organizational tolerance for risk. Determining a risk-relevant relationship that reflects the effect of security risk management uncertainty is the crux of the problem.

3.3 Risk-Relevant Time and Distance Scales

All measurable objects and phenomena have characteristic time and/or distance scales. The formation of the universe occurred over billions of years. Bacterial colonies evolve via mutations expressed within days, hours or even minutes. Chemical reactions occur on sub-second time scales. Astronomers observe objects that are many light-years away and atomic physicists study atoms that are 0.1 to 0.5 nm (10^{-9} m) in size.

The so-called "Multiverse," a hypothetical group of universes comprising everything that exists, spans a distance scale corresponding to 60 orders of magnitude; from the Planck scale (10^{-35} m) to the Hubble scale (10^{25} m). Interested readers are encouraged to visit the planetarium at the American Museum of Natural History in New York City to view an impressive display in the planetarium that depicts the physical scale of objects in the Universe.

IT environments also have characteristic time and distance scales. The time scales are dictated by multiple factors, which include the business activities supported by a given environment, the longevity of security controls and/or the duration of processes implemented via technology. Distance scales in IT environments relate to either the physical or virtual presence of information assets, broadly defined.

The risk-relevance of time and distance scales in assessing cybersecurity risk is not fully appreciated. This situation exists despite the fact that many risk-relevant activities are linked to such scales, most notably the application of security controls to risk factors, i.e., security risk management.

For example, it would make little sense to make quotidian updates to an organization's information security policy since risk-relevant changes to the risk profile typically occur over longer time scales. Similarly, it would not be productive to conduct monthly security reviews of a business process if that process remains relatively constant over year-long time scales and other risk-relevant issues are similarly static.

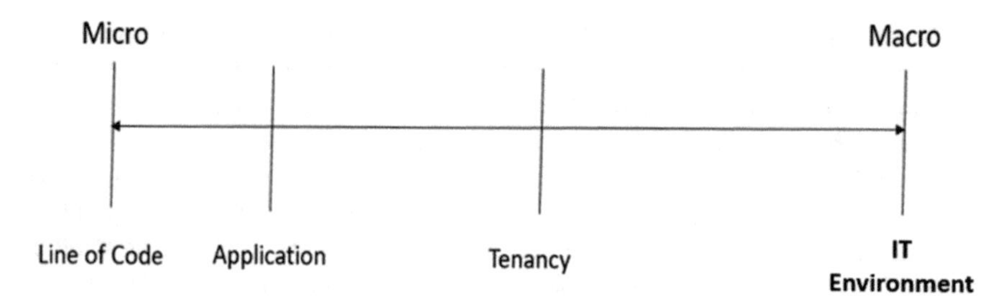

Fig. 3.2 An indicative IT environment "distance" scale

To recap, examining software at the code level is required to prevent attacks such as cross-site scripting. Conversely, reviewing software for code-level vulnerabilities will not identify issues that are only manifest on an enterprise scale. The upshot is the scale used to assess cybersecurity risk must align with the scale of the risk-relevant phenomenon, process or vulnerability being assessed.

Figure 3.2 illustrates a conceptual "distance" scale of a generic IT environment. It specifies objects of cybersecurity risk assessments sequenced according to the virtual "space" they occupy.

Of course, Fig. 3.2 is merely indicative. It is impossible to specify exact "distances" in an IT environment. In fact, the very notion of distance in the virtual world doesn't coincide neatly with the physical world. This incongruity is of little consequence. It is clear these entities have an increasingly broader if virtual cross section and therefore require an increasingly broader perspective to assess.

A digression on statistics is a useful prelude to discussing the concept of correlation-time pursuant to a more fulsome appreciation for risk-relevant time scales.[1] In addition, the discussion is useful more generally since these concepts are fundamental to statistics.

The statistics associated with measurements of some continuous variable R are typically described in terms of the moments of the probability distribution of R, i.e., P(R). P(R) characterizes the probability of observing R between R and R + dR, where dR is an infinitesimal fraction of R. The average and mean square value of R are defined as follows, where the integral is a continuous summation with respect to the variable R,

$$\text{Average of R} = \langle \mathrm{R} \rangle = \int \mathrm{RP(R)dR}$$

$$\text{Mean Square Value of R} = \langle \mathrm{R}^2 \rangle = \int \mathrm{R}^2\mathrm{P(R)dR}$$

The average and mean square value can also be written as discrete averages, designated by brackets, which are based on a large number of measurements (N) of the values of R as follows:

[1] Quantum Chemistry Course Notes (5.74), MIT Open Courses, Department of Chemistry (2009).

$$\langle R \rangle = (1/N) \sum R_i, \text{ where i ranges from 1 to N}$$

$$\langle R^2 \rangle = (1/N) \sum R_i^2, \text{ where i ranges from 1 to N}$$

The variance (σ^2) of the probability distribution P(R) measures the deviation from the average value and is defined as follows,

$$\sigma^2 = \langle R^2 \rangle - \langle R \rangle^2$$

The square root of the variance equals the standard deviation of the distribution. Both specify the dispersion or uncertainty about the average or mean value of P(R).

We can also describe the statistical relationship between two variables by defining a *joint* probability distribution P(R,Q) for two variables P and Q. P(R,Q) characterizes the probability of observing R between R and dR *and* the probability of observing Q between Q and dQ.

As with the probability distribution of a single variable, the moments also describe the statistics of a joint probability distribution. However, in this case the moments reveal the statistical relationship between the two variables R and Q. Therefore, the variance is now represented by the covariance (C_{RQ}) i.e., the variance of a bivariate distribution, which is defined as follows,

$$C_{RQ} = \langle RQ \rangle - \langle R \rangle \langle Q \rangle$$

C_{RQ} is a measure of the correlation between the variables R and Q. Qualitatively, it measures whether a choice of specific values of R implies the associated values of Q have different statistics from the values of R. In other words, if the variables R and Q depend the same way on a common internal variable, they are correlated. The variables are uncorrelated if no statistical relationship exists between the two variables.

The covariance is applicable to any two continuous variables. In the real world it is frequently used to describe variables in time and space. However, instead of two internal variables, our interest is in evaluating the same variable at different points in time, i.e., the *autocorrelation* between R evaluated at time equals zero and at some later time t.

Using this statistical machinery we can specify a highly idealized but nevertheless instructive method of determining a risk-relevant time scale for IT environments. This idealization is based on two assumptions: risk factors are random variables and the aggregate effect of all risk factors can be represented by an average value that fluctuates in time. The points of this exercise are to showcase the power of stochastic processes in this context and to illustrate the relevance of identifying a characteristic time scale for IT environments.

If we assume the potential for information compromise is described by a single random variable R, we want to know the correlation between subsequent measured values of R as a function of time. To reiterate, the value of R represents the (fictitious) aggregate effect of all risk factors in an IT environment. The measured values of R

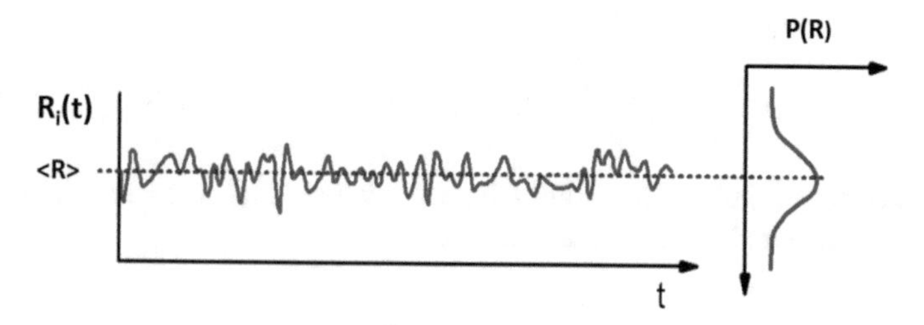

Fig. 3.3 Fluctuations of R(t) about the mean value (<R>)

($R_i(t)$) are assumed to fluctuate randomly about some mean value as shown in Fig. 3.3, and these fluctuations result in a probability distribution P(R).

We wish to evaluate the ensemble average of R using the time-correlation function C(t), which is defined as follows,[2]

$$C(t) = \langle R(0)R(t) \rangle \tag{3.1}$$

Here R(0) and R(t) are the values of the potential for information compromise across the IT environment measured at time $t = 0$ and some later time t. Recall R is a single variable representing the resultant potential for information compromise arising from the aggregate effect of *all* risk factors in the IT environment. C(t) represents a running tab of the autocorrelation between R(0) and subsequent values of R, i.e., R(t).

C(t) can be normalized to unity by dividing by <R(0)R(0)>. Since R(0) is perfectly correlated with itself, the normalized value of C(t) reveals the decay of the IT environment from perfect correlation, i.e., $C(t) = 1$, as time progresses. The short-time value of C(t) is proportional to the average of "R-squared" = $\langle R^2 \rangle$ whereas the asymptotic, long-time value is proportional to the average of "R"-squared = $\langle R \rangle^2$ as shown in Fig. 3.4.

In other words, at some time $t > 0$, C(t) devolves from its initial normalized value of 1 and asymptotically approaches the steady-state value of zero. The long-term value of $C(t) = \langle R \rangle^2$. The point is the long-term value of C(t) is statistically unrelated to its initial value measured at time $t = 0$, i.e., $\langle R^2 \rangle$. The time scale for scenario self-similarity would be determined by the decorrelation decay time.

In other words, C(t) would establish the time scale for valid risk assessments; the aggregate value of IT environment risk factors is no longer statistically related to a previously measured value. Therefore, results of future assessments based on historical conditions that exceed the correlation-time would be invalid. Presumably C(t) would decay rapidly in IT environments given the frequency and number of risk-relevant activities such as authentications of identity, access to web sites, clicking on embedded hyperlinks, etc.

[2] Ensemble and time averages are discussed in Chapter 6.

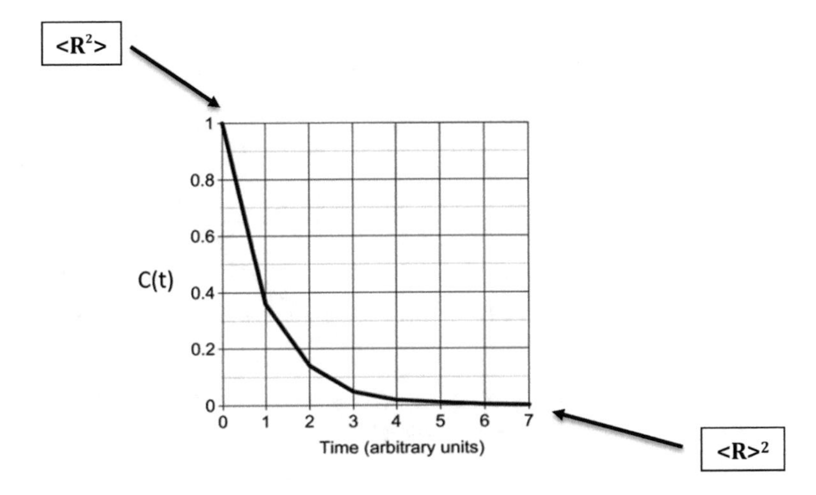

Fig. 3.4 Hypothetical IT environment correlation-time

Of course and as noted previously this formulation represents an idealized scenario. Identifying a single variable in an IT environment that reflects the overall magnitude of cybersecurity risk is fanciful. Specifying its time series would likely be even more problematic. The identified variable would be the cybersecurity-equivalent of the Dow Jones Industrial Average plotted as a function of time.

3.4 Power Laws and Scaling Relations

If the independent variable of a function contains an exponent other than unity, the dependent variable is said to obey a *power law*. Such a function is a specific type of scaling relation. Examples of such functions include $y = x^2$, $y = x^3$, etc. Often the independent variable of a scaling relation describing a risk-relevant phenomenon relates to distance or time.

The exponent of a scaling relation governs the rate of change of the dependent variable as the independent variable varies over some defined range. If the two variables actually describe a risk-relevant relationship within an IT environment, the exponent will reveal the precise nature of that relationship, which in turn could affect the application of security controls.

The axes of a graph depicting a scaling relation must be scaled according to each parameter's range of values. As noted above, the scale of each axis is analogous to the scale of a map, where incremental distances along an axis will determine the observable detail. If the distance between unit increments is too large, risk-relevant details might not be visible. If such increments are too small, the full range of parameter values might not be represented with attendant consequences to the assessment results.

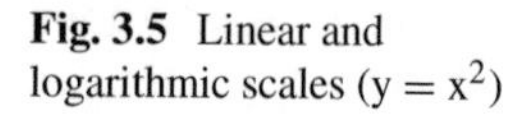

Fig. 3.5 Linear and logarithmic scales ($y = x^2$)

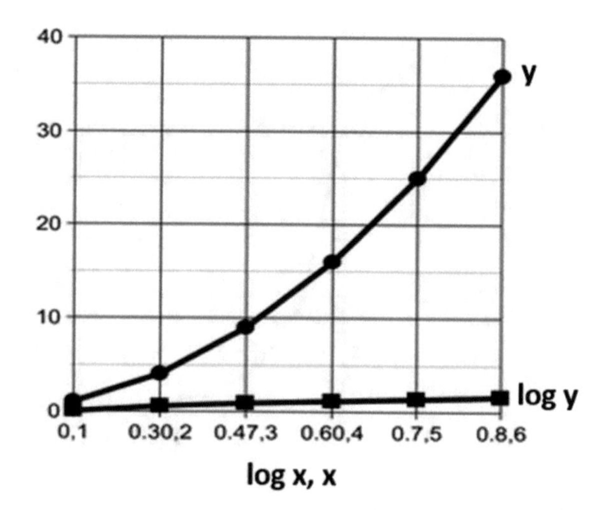

If a linear scale is used to plot a trend line it can result in insufficient "real estate" especially if the independent variable has an exponent greater than one. In such instances a logarithmic scale can be used along one or both axes. Since the integer values of the logarithm (base 10) represent incremental powers of ten, i.e., 10^1, 10^2, 10^3, etc., this representation is more compact.

Figure 3.5 displays the function $y = x^2$ using both linear and logarithmic scales, where the rectangles and circles correspond to the values of the logarithmic and linear plots, respectively. The non-linear behavior of the function is strikingly apparent using a linear scale. In contrast, the function appears to have a constant slope when viewed on a logarithmic scale. Aging readers might consider reporting their age using a logarithmic scale. The vast majority of readers will report a figure far less than two, thereby experiencing a modest if illusory sense of rejuvenation.

To repeat for emphasis, the scale used to display each axis affects the observable detail. The price for revealing more detail is the need to delineate a broader range of values of the independent and/or dependent variables.

The linear trend line in Fig. 3.5 becomes increasingly curved for larger values of x. This curvature cannot be seen using the logarithmic scale, and therefore might not provide sufficient granularity depending on the requirement. For functions obeying power laws with even bigger exponents, e.g., $y = x^4$, $y = x^5$, etc., the effect would become increasingly pronounced.

The concept of *scale-invariance* is relevant to functions obeying power laws, and therefore can have implications to various risk-relevant phenomena.[3] Specifically, scaling the argument of a function f(x) by a constant factor (c) causes a proportionate scaling of the function itself. That is,

$$f(cx) = (cx)^{-k} = c^{-k}f(x) \sim f(x)$$

[3] https://en.wikipedia.org/wiki/Power_law.

In other words, scaling the argument by a constant factor simply multiplies the original power law by the constant c^{-k}. The implication is that all power laws with the same exponent are equivalent up to constant factors since each function is simply scaled-up or scaled-down versions of the others.

Computer networks can demonstrate scale invariance. This condition occurs when the degree of connectedness obeys a power law and is therefore scale-invariant. In other words, the probability p(k) of nodes in the network having k connections to other nodes is proportionate to $k^{-\gamma}$. The exponent γ has been shown to be relevant to virus propagation and persistence in computer networks.[4]

Finally, we note that the scale invariance of power laws reveals the linear relationship between log f(x) and log(x), where the slope of the resulting line is given by the power law exponent.

3.5 The *Power* of Scaling Relations

The tongue-in-cheek title notwithstanding, a power law scaling relation between risk-relevant variables can provide insight into the magnitude of risk and/or the effects of security controls. Although power laws in cybersecurity are rare, they do exist. The relationship between entropy and password complexity is one notable example, where there is an exponential relationship between password complexity and password cracking time.

The following discussion focuses on one particular power law scaling relation that relates to WiFi threat scenarios. It is unique within cybersecurity because it derives from physical principles. Details are presented in order to demonstrate the risk-relevant implications of a power law relationship in this context as well as the relevance of scaling relations more generally.

Wireless access points radiate electromagnetic energy in the form of a radiofrequency signal that typically oscillates at a frequency of either 2.4 or 5 GHz.[5] WiFi signals are notoriously promiscuous by design. That is, anyone possessing the appropriate technology can detect the emanating radiation assuming the signal has sufficient intensity relative to the ambient noise power at the point of detection.

The detection of WiFi signals by unauthorized individuals enhances the potential for compromising the information contained therein. Adversaries frequently use man-in-the-middle attacks to carry out these information compromises and such attacks are well documented.[6] Encryption does not address all threat scenarios, and

[4] D. Chang and C. Young, *Infection Dynamics on the Internet*, Computers and Security, 24, 280–286, 2005.

[5] A gigahertz (GHz) equals 10^9 Hz (Hz), i.e., oscillations-per-second. A megahertz (MHz) equals 10^6 Hz.

[6] What Is a Man-in-the-Middle Attack? https://us.norton.com/internetsecurity-wifi-what-is-a-man-in-the-middle-attack.html.

therefore unauthorized signal detection is considered a risk factor for information compromise.

We note that although this scenario is describing the likelihood component of risk, one might arguably contend it is describing vulnerability. This ambiguity results from the nature of signals containing information. Specifically, if the signal is compromised, the information contained therein is lost.

Therefore, in determining the potential for signal interception we are simultaneously estimating the loss resulting from such a compromise. In other words, it is an all or nothing-at-all proposition. The same argument could be posed for other cyber-attacks where assessing the potential for information compromise simultaneously represents an assessment of the vulnerability component of risk.

A specific power law governs the intensity of radiating WiFi signal energy as a function of the distance from the radiating source. Specifically, the signal intensity, i.e., the radiated power per unit area, scales inversely with the square of the distance between the transmitter and receiver for point sources of radiation. WiFi access points qualify as point sources because the wavelength of radiated energy is typically much less than the distance between the access point and the point of signal reception.[7]

The critical parameter affecting authorized and unauthorized signal detection is the signal power-to-noise power ratio at the point of reception. Note that signal encryption presumably prevents message comprehension but it can do nothing to prevent unauthorized signal detection.

That said, encrypting data in transit and at rest is always recommended. Current WiFi standards call for WPA2 encryption, which is a National Institute of Science and Technology (NIST) FIPS 140-2-compliant, AES encryption algorithm. Current WiFi systems use 802.1x-based authentication as of this writing.

The signal-to-noise ratio at the point of reception exclusively determines if a signal is detectible. This ratio is a function of three parameters as measured at the reception point: the distance from the radiating source, the ambient noise power, and the sensitivity of the detection equipment.

Consider a WiFi access point with a transmit power (P_t) of 100 milliwatts (mW) radiating omni-directionally, i.e., zero gain, at a frequency of 2.4 GHz.[8] At this frequency the signal wavelength (λ) is 12.5 cm (0.125 m) in air. As noted above, the access point is assumed to be a point source, which is radiating energy into an imaginary sphere whose surface area scales as the square of its radius. Therefore, the intensity or power density of the radiating signal *decreases* inversely with the square of the sphere radius.

If the receiver antenna is assumed to be a dipole located 10 m from the access point with an approximate cross section equal to $\lambda^2/4\pi$, the signal intensity at the

[7] The 802.11 WiFi standard encompasses multiple signal frequencies to include 900 MHz, 2.4 GHz, 3.6 GHz, 4.9 GHz, 5.9 GHz, 6 GHz and 60 GHz. The wavelength of a 2.4 GHz WiFi signal is 12.5 cm. The wavelength of electromagnetic signals in non-dispersive media scales inversely with frequency according to the following relationship: wavelength = speed of light/frequency. Therefore, it is possible to calculate all other wavelengths from a single wavelength/frequency pair. For example, the wavelength of a 5.9 GHz signal equals (2.4/5.9) × 12.5 cm ~5.1 cm.

[8] A milliwatt (mW) equals one thousandth of a watt.

receive antenna (P_r) equals the radiating signal power density at that location times the antenna cross section, i.e., the effective area of the antenna as "seen" by the signal. Therefore, the power at the receiver (P_r) equals,

$$\begin{aligned} P_r &= P_t/4\pi r^2 \times \lambda^2/4\pi \\ &= \left(0.10\,\text{watts}/100\,\text{m}^2\right) \times \left(0.016\,\text{m}^2/16\pi^2\right) \\ &= 0.001\,\text{W/m}^2 \times \left(0.016\,\text{m}^2/157.955\right) \\ &= 0.001\,\text{W/m}^2 \times 0.0001\,\text{m}^2 \\ &= 0.1 \times 10^{-6}\,\text{W, i.e., } 1/10\text{th of a microwatt} \end{aligned}$$

This signal power might not seem particularly robust but modern radio receivers can easily detect much weaker signals. Therefore, receiver sensitivity would not be an issue in this example.

In general, the ratio of receive power, P_r to transmit power, P_t, is given by the Friis transmission formula[9]:

$$P_r/P_t = G_r G_t (\lambda/4\pi R)^2 \tag{3.2}$$

The gain of the receiving and transmitting antennae, G_r and G_t, are functions of their respective efficiency and directivity relative to the source.[10]

Rearranging (3.2) yields the received signal power, P_r at a distance R from the transmitter source. Recall P_t equals 100 mW, and G_r, G_t and R can be measured. Therefore, every quantity needed to determine the potential for successful signal interception is known except for the ambient noise power, which can either be measured or estimated from published data.[11]

Figure 3.6 depicts a radiating point source and the inverse-square power law dependence of signal intensity with distance.

What does this power law relationship between intensity and distance imply? If the signal intensity is measured to be I at a distance R, the intensity at a distance 2R becomes proportional to $I/(2R)^2$. At a distance 4R the signal intensity becomes proportional to $I/(4R)^2$, etc. Therefore, the R^{-2} power law provides a recipe for determining the potential for unauthorized signal detection *at any distance from the source.*

Figure 3.7 shows the non-linear/inverse-squared dependence of intensity with distance using arbitrary units for both distance and intensity. The signal intensity is normalized to unity at the transmitter location.

---

[9] Shaw, J., Radiometry and the Friis transmission equation, American Journal of Physics (1), January 2013.

[10] Directivity describes how much of an antenna's radiated power is transmitted in a given direction (per solid angle) relative to the amount radiated by an isotropic antenna radiating uniformly into 4π steradians (Sr).

[11] Test Operations Procedure; 02-6-595 Characterization of an Outdoor Ambient Radiofrequency Environment, www.apps.dtic.mil.

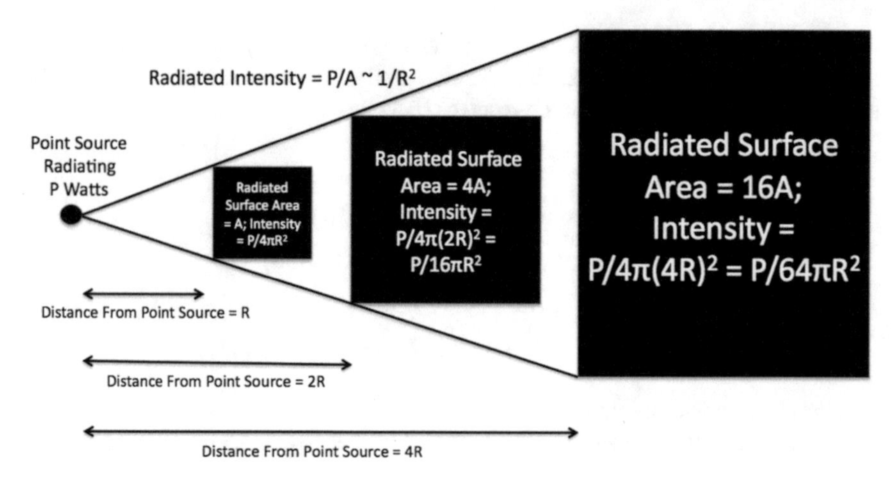

Fig. 3.6 A point source of radiation

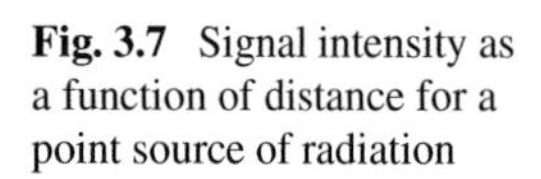

Fig. 3.7 Signal intensity as a function of distance for a point source of radiation

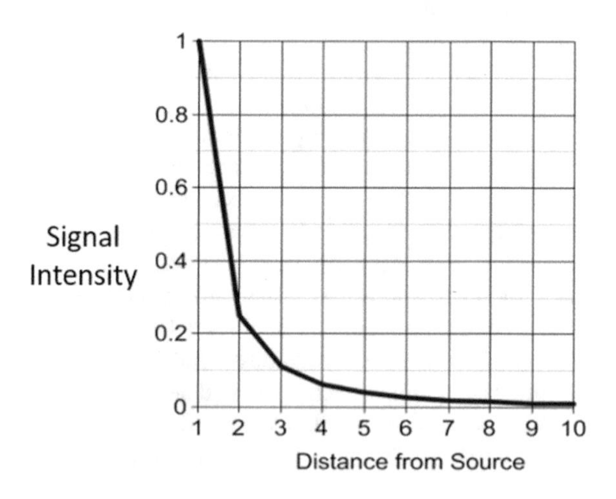

The non-linearity of the scaling relation has an operationally important implication to both attackers and defenders, which can be summarized in a single phrase: a little goes a long way. In other words, a small change in the distance between the receiver and transmitter will have a disproportionate effect on the signal-to-noise ratio and hence the potential for successful signal interception.

Therefore, physically extending the perimeter to exclude an attacker's physical presence will go a long way to reducing the potential for signal detection and perhaps the potential for information compromise. Conversely, if an attacker could inch their way closer to the radiating access point the potential for success increase disproportionately with distance.

Note that the signal-to-noise ratio only scales linearly with signal intensity. Therefore, reducing signal intensity as a defensive strategy only has a linear effect on the potential for successful signal interception. This observation has important operational implications, and highlights the criticality of identifying linear versus non-linear relationships between risk-relevant variables. The salient point is understanding the precise dependencies relative to the potential for information compromise would not be possible without a proper scaling relation between risk-relevant parameters.

As noted above, (3.2) enables generalizations about the potential for unauthorized signal detection for *any* threat scenario where similar conditions apply. In this instance "similar" means a point source of electromagnetic radiation. The ultimate implication of this scaling relation is that the distance between the access point and an attacker's receiver is a risk factor for information compromise in WiFi threat scenarios.

An adversary and an authorized network user attempting to demodulate a WiFi signal must be able to detect sufficient radiated power relative to the ambient noise power at their respective points of signal reception. Physics does not discriminate based on ethics. Therefore, the signal-to-noise power is a *bona fide* metric for *both* system performance and the potential for information compromise. It represents the minimum threshold for signal detection for both legitimate WiFi users and adversaries. The minimum signal-to-noise ratio required for signal demodulation is often quoted to be about 10 decibels (dB).[12]

A scaling relation is powerful because it expresses a precise relationship between risk-relevant variables. The scaling relation in the WiFi threat scenario is a power law, which is scale-invariant. Therefore, the intensity versus distance power law can simply be scaled up by a constant factor if scenario-specific conditions change.

The lesson is scaling relations can provide insight into risk-relevant dependencies that can guide the application of security controls to risk factors. However, their interpretation and associated implications with respect to the creation of a risk management strategy must be tempered by the assumptions used to establish such relations.

The physical dependencies predicted by (3.2) are exact and experimentally verifiable. As noted above, the measured signal-to-noise power ratio represents an unambiguous metric for unauthorized signal detection. Therefore, this metric actually expresses *absolute risk* because the relation is derived from immutable physical principles that can be directly related to the potential for information compromise.

In contrast, other scaling relations might be based on assumptions that deviate from real world conditions. The accuracy and hence operational applicability of results so derived might be vitiated by such assumptions. However, such scaling relations could possibly identify *relative* risk and therefore might be used to compare threat scenarios. These scaling relations might be no less meaningful, but the implications might appreciably differ from those based on measurable quantities.

---

[12] For two quantities R_1 and R_2 a decibel is defined as 10 log (R_1/R_2). The decibel is frequently used to express relative differences in amplitude. If power is proportional to $(R_1/R_2)^2$, relative differences in R_2 and R_1 can be expressed as 10 log $(R_1/R_2)^2$ or equivalently 20 log (R_1/R_2).

The bottom line is all scaling relations are not created equal. Care must be exercised in interpreting their results and in declaring any operational implications.

3.6 Authentication and Scale

The potential for information compromise is of particular concern when authenticating a network user's identity. Unauthorized access to an IT environment resource can result in the disclosure of all information to which the legitimate network user has access.

Recognition of this condition plus the ease of password compromise due to credential theft and/or social engineering has resulted in the widespread adoption of multi-factor authentication. Multifactor authentication uses multiple credentials communicated via independent channels to authenticate the identity of an individual requesting electronic access to an information asset. The tacit assumption is that it is highly unlikely an attacker could simultaneously compromise more than one channel and thereby steal the credentials necessary to achieve unauthorized access.

The number of authentications scales linearly with the number of number of systems/applications requiring authentication assuming a constant number of users. For example, if an IT environment has N network users, and each user must authenticate separately to M distinct applications, the maximum number of authentications is given by N $\times$ M. As noted previously, the contention is each instance of identity authentication is a risk factor for information compromise.

Figure 3.8 plots the number of authentications as a function of the number of applications for 100 network users.

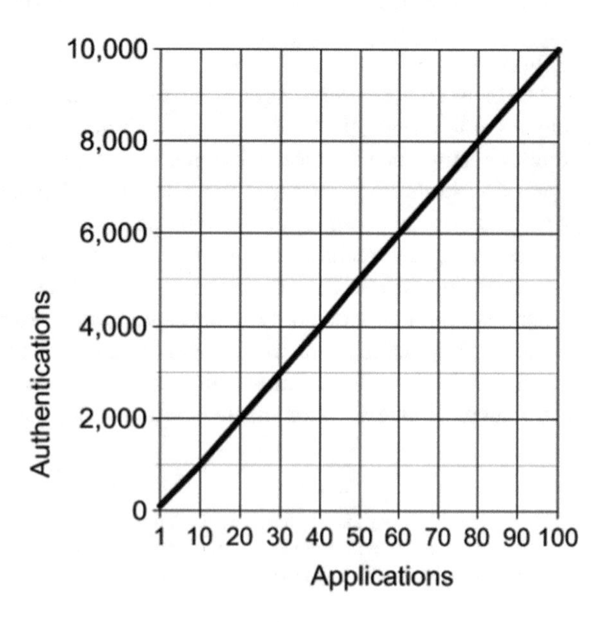

Fig. 3.8 Authentications as a function of the number of applications (100 network users)

Since the number of network users and applications are not easily adjusted, one risk management strategy would be to facilitate access to all applications to which an individual has access privileges once proper authentication has occurred. This approach would obviate the need for multiple authentications each time a network user logs-in. Such an approach would potentially have a significant effect on the total number of authentications.

To that end, deploying a so-called single sign-on solution has the desired effect, where the number of authentications equals the number of network users. For example, if an IT environment consists of N users, the number of authentications is now independent of the number of applications and is just a function of N. In other words, the number of authentications depends only on N rather than $N \times M$, where M is the number of applications. The effect of single sign-on solutions on the potential for information compromise will be analyzed in more detail in Chap. 10.

In general, security controls that reduce the number of risk-relevant processes can have a dramatic effect on reducing cybersecurity risk. The reason this reduction is so effective is because it decreases the number of risk factors, which we will see has a non-linear effect on the potential for information compromise.

Chapter 4
IT Environment Dimensions and Risk Factors

4.1 Introduction

Risk factors for information compromise are prevalent throughout IT environments. Information technologies as well as processes that rely on these technologies have features that can be exploited by individuals with malicious intent. Some of these risk factors also facilitate unintentional data leakage. The autocomplete function in email is one well-known example of this phenomenon.

An IT environment can be viewed as a physical and virtual ecosystem consisting of multiple dimensions relating to information technology, information management, and network users. Risk factors for information compromise exist in all dimensions, and some risk factors can be present in more than one dimension. Network user behavior generates specific risk factors for information compromise.

It is instructive to compare IT and physical environments with respect to their security risk profiles. Physical environments divide neatly into internal and external areas, where external areas are contiguous with the outside environment. Each area has a physical portal that facilitates pedestrian ingress and egress.

Verifying identity and physical access privilege is generally straightforward since building portals are finite in number, visible and manageable via centralized access control systems. The vulnerability and impact components of risk associated with unauthorized physical access are significant, but the potential for a successful breach is typically low if appropriate security controls are implemented.

Obviously a physical presence is required to achieve physical access within a given space. The floor layout dictates the sequence of portals necessary to gain access to internal areas. The floor plan also remains relatively constant over risk-relevant time scales. The deployment of security controls will at least partly depend on the portal sequencing, and security risk management resources can be concentrated to maximally reduce the potential for unauthorized entry.

It is significant that theft in the physical world requires the physical removal of the pilfered items. The perpetrators can theoretically disguise themselves to hide their identity but they cannot make themselves invisible, at least not as of this writing.

C. S. Young, *Cybercomplexity*, Advanced Sciences and Technologies for Security Applications, https://doi.org/10.1007/978-3-031-06994-9_4

This constraint limits the opportunities for thieves since they must physically abscond with the booty. Less opportunities for physical theft translates to a reduction in the number of risk factors for unauthorized physical access to restricted space.

In summary, physical security risk factors are limited, easily identified, and do not vary appreciably over risk-relevant time scales. Opportunities for theft are constrained by the physical design of the building, and the potential for a successful breach is reduced by a strategic concentration of security controls. In general, security risk management is prescriptive and threat scenario variations are limited.

IT environments present a very different risk profile. First, the identity of individuals accessing IT environments from the outside world cannot be authenticated using sensory perception, e.g., vision, hearing. Second, network users routinely communicate with individuals outside their organization and who are located beyond the perimeter firewall. Third, information compromises do not require the information to be missing. Therefore, physical access to items targeted for compromise is not required. Finally, the number of network elements, information assets and associated risk factors for information compromise are vast in comparison with typical physical security environments, and the internal workings of IT environments are largely opaque to network users.

One might reflexively suggest that cybersecurity threat are more *complex* than their physical security counterparts. If asked to be more specific one might respond by saying the difference in complexity is due to both the sheer number of risk factors and the uncertainty associated with risk management in each type of scenario.

Intuitively this answer feels right, and it represents a good start to identifying a more precise description of complexity in IT environments. If risk factors are responsible for increases in the magnitude of one or more components of risk, differences in the *number* of risk factors should have some effect on the potential for compromising affected assets.

The relatively prescribed and physically bounded physical security threat scenarios are correspondingly limited in their variability. They are relatively fixed in space and time and are therefore amenable to physical surveillance. In addition, risk-relevant activities such as identity verification/access validation and subsequent entry via turnstiles typically occur in series.[1] In contrast, numerous network users engage with risk-relevant systems in parallel. Therefore, IT and physical environments can also be distinguished by their respective differences in risk management uncertainty.

Although intuition is a helpful adjunct to formal assessments of risk, more rigor is required if the hope is to provide a general description of complexity and thereby apply it to any IT environment. Security risk management uncertainty and the number of risk factors for information compromise appear to be promising candidates for the

[1] It is more precise to say risk-relevant activities associated with physical security threat scenarios are frequently linear. For example, organizations routinely utilize multiple turnstiles in parallel, but the simultaneous throughput and hence the risk-relevant activity scales linearly with the number of turnstiles.

most significant drivers of complexity in IT environments. However, we really need to know their specific contribution to complexity in this context.

The ultimate objective is to identify a model for complexity in IT environments. If the model is accurate, or at a minimum provides useful insights, it should also confirm intuition with respect to the drivers of complexity. Differences in physical and IT environments have hinted at these drivers, but a model would provide the required generality.

4.2 Information Management

IT environments can be described in terms of multiple dimensions that contain risk factors for information compromise and/or information-related business disruption. Here we are speaking of electronic information, which exists as ones and zeros in computer memory. Information is not an abstraction. Both physical and electronic forms of "traditional" information must be located somewhere, must relate to some topic(s), and must have a lifecycle.

Figure 4.1 depicts a "vector" representation of information management within what we refer to as "Information Management 3-Space." The notion of vectors in 3-space is borrowed from physics. It is appealing because it graphically shows the multi-faceted nature of information, where the magnitude of the risk of information compromise can vary by dimension.

Specifically, a unit of information is represented as a vector with spatial (information source), temporal (information lifecycle) and contextual (information type)

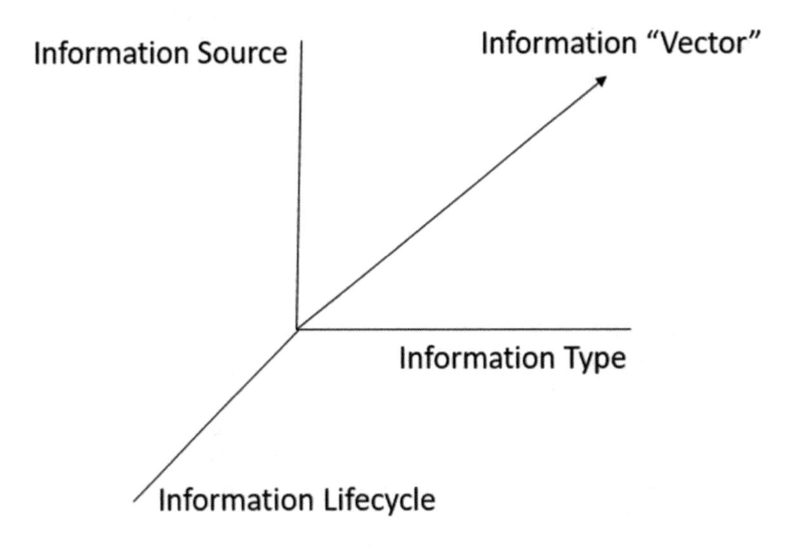

Fig. 4.1 Information vector in information management 3-space

Table 4.1 Information management dimension elements

Information source	Information type	Information process/lifecycle
Email	Business-strategic	Creation
Servers/endpoints	Human resources	Electronic transmission
Messaging services	Client-confidential	Physical transport
Mobile devices	Technical	Reproduction/replication
Social media	Vendor-3rd party	Retention/storage
Physical media	Health	Electronic deletion
Cloud applications	Salary	Physical destruction

components. These dimensions are somehow superimposed on the multi-faceted elements of IT environments, where information typically exists within such environments.

Notably, risk factors for information compromise affect all three information management dimensions. In addition, risk factors can simultaneously relate to more than one dimension. The result is an increase in the risk factor count without adding elements to the IT environment. In addition, relationships between information management risk factors can affect risk management uncertainty.

Table 4.1 lists the three information management dimensions and examples of elements within each dimension.

Furthermore, each element can present unique risk factors for information compromise, and these risk factors can affect different components of risk. Multiple information dimensions and varied components of risk result in a cornucopia of risk factors that might require a spectrum of security controls.

Since information technology is required to manage electronic information from creation to destruction, and network users continuously interact with this technology, information technology and network users represent two additional dimensions that contain risk factors for information compromise.

4.3 Information Technology

All IT environments contain information technologies with risk factors that can potentially be exploited to facilitate unauthorized access to information assets. IT networks are themselves multi-faceted whose functionality is described by the Open Systems Interconnection (OSI) Model. The OSI Model is a conceptual framework used to describe the disparate functions of a networking system. Specifically, "It reveals the multi-layered structure of IT systems, and establishes a universal set of rules and requirements that support interoperability between varied software and hardware products."[2]

[2] https://en.wikipedia.org/wiki/OSI_model.

The OSI model characterizes any networked computer or telecommunication system irrespective of the specific products contained therein. It also reveals how the layers interact to facilitate networked communications. The model will not be reviewed in detail, but the seven layers are listed below and are accompanied by a brief description.[3,4]

Layer 1: Physical Layer-Transmits raw bit streams over physical media, e.g., wires and fibers.

Layer 2: Data Link Layer-Defines the format of the data on the network.

Layer 3: Network Layer-Determines the data path.

Layer 4: Transport Layer-Transmits data using transmission protocols, e.g., TCP, UDP.

Layer 5: Session Layer-Maintains connections and controls ports and sessions.

Layer 6: Presentation Layer-Ensures data is in a usable format and is where data encryption occurs.

Layer 7: Application Layer-Human–computer interaction layer where applications can access network services.

The layers are actually abstractions that are convenient representations of the elements within Internet Protocol (IP)-based environments. Significantly, risk factors for information compromise can be present in each of the OSI layers. Notably, information technology and information management are clearly not independent, which can contribute to security risk management uncertainty.

Imagine attempting to transmit packets using an IP-based system without a wire or fiber (Physical Layer 1). Equally, sending an email without a native email client or via a webmail application (Application Layer 7) is simply not possible. In other words, each layer contains one or more IT environment elements that are necessary to process electronic information.

Moreover, specific combinations of risk factors present in different layers can enhance the potential for information compromise. For example, use of a low-complexity password to protect an information asset accessible via a hidden path to the internet in an IT environment with an open network architecture enhances the potential for information compromise. The point is this specific combination of risk factors could elevate the potential for information compromise relative to many individual and largely unrelated risk factors.

Figure 4.2 depicts the information management and information technology dimensions of an IT environment.

Finally and as noted above, risk factors can affect more than one component of risk in addition to multiple IT environment dimensions. The net result can be a multi-fold increase in the number of risk factors and/or an increase in risk management uncertainty. Figure 4.3 graphically illustrates this effect, where the letters L, V and I refer to the likelihood, vulnerability and impact components or risk, respectively.

[3] www.imperva.com.

[4] Arguably there are eight OSI layers, where the eighth layer corresponds to network users.

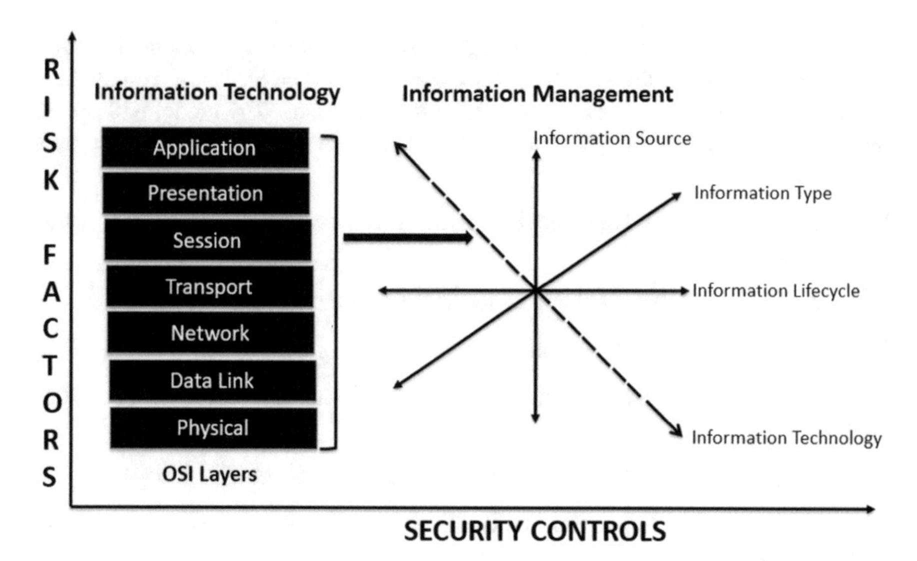

Fig. 4.2 Multi-dimensional IT environments

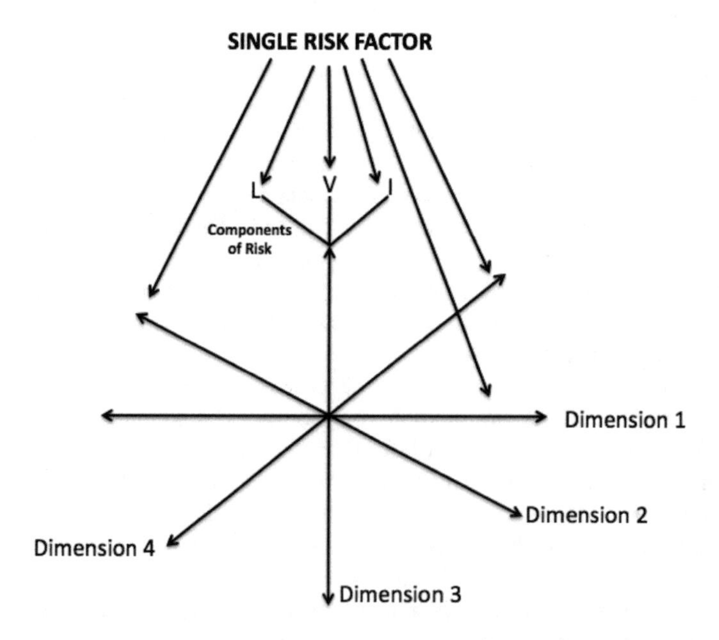

Fig. 4.3 Single risk factor affecting multiple IT dimensions and multiple components of risk

4.4 Network Users

Perhaps the least appreciated but most impactful area of cybersecurity risk management relates to network users. These individuals are responsible for all actions associated with the information lifecycle and are constantly interacting with information technologies. Network users are as much a part of the IT environment ecosystem as information technology, and they actually affect technology performance. In contrast with other technologically-laden environments, network users are not merely passive elements.

Numerous books have been written about human behavior so nothing original about that topic will be stated here. However, it is worth recalling the potential for information compromise is frequently rooted in the imbalance between the need for security and the desire for convenience. The latter is a characteristic of human behavior and is perhaps a trait inherent to our species.

There is no escaping the fact that network users contribute to the potential for information compromise. They are inextricably linked to the processes, workflows and technologies that are integral to managing information. Moreover, they directly interface with networked devices that process information, where the operational word here is "networked." Therefore, network users influence the specific devices they operate but also affect entities across the ecosystem.

Managing human behavior is integral to a comprehensive cybersecurity risk management strategy. A machine or application is agnostic to the intent of the individual inputting instructions. Technologies do not spontaneously generate risk factors for information compromise. Computers dutifully respond to the commands of network users, for better or for worse.

However, humans are also the reason IT environments exist. Although it would be naïve to suggest human behavior does not contribute to the potential for information compromise, it would be inaccurate and perhaps unjust to claim humans are the exclusive cause of cybersecurity incidents.

Clearly *something* causes information compromises. The point is the finger cannot be pointed at a particular entity in isolation. Humans utilizing processes, workflows and technologies relating to information management somehow create conditions within IT environments that increase one or more components of risk with respect to the potential for information compromise. Information technologies contain risk factors for information compromise and human behavior is sometimes weaponized to exploit these vulnerabilities.

Focusing on specific individuals and/or technologies is clearly required but not to the exclusion of the big picture. If IT environments were as prescribed as physical environments, a vigilant eyeball could simultaneously be kept on all risk factors that contribute significantly to the potential for information compromise. If a *macroscopic* cybersecurity risk management strategy is being implemented there are certainly frequent lapses. The reason for those lapses is likely because there are too many risk factors and/or the status of any particular risk factor (or combinations thereof) cannot be precisely pinpointed as a function of time.

The need for a macroscopic view of cybersecurity risk suggests imposing as much uniformity across the environment as possible. All IT environments actually invoke such a strategy, but the motivation is more likely to achieve economies of scale rather than security risk management scaling. For example, IT environments typically utilize only one email application, one word processing application, one electronic spread sheet, etc.

Such a configuration results in minimum security risk management uncertainty since it minimizes deviations from the canonical set-up. It also enables addressing the maximum number of risk factors for information compromise using the minimum number of security controls with attendant resource implications.

One might argue that a vulnerability common to a single application would be replicated across the number of software installations in the environment. The counter argument is vulnerability fixes work the same way: disparate applications with disparate vulnerabilities would require disparate fixes assuming the spectrum of vulnerabilities could be identified.

The efficiency and effectiveness of such a strategy is itself compelling but we must do better to tackle complexity in IT environments. Whatever the IT environment configuration, the aggregate effect of risk factors and security risk management uncertainty clearly increase as the scale of the environment widens. The inescapable conclusion is the magnitude of cybersecurity risk grows in kind, and therefore the scale of a cybersecurity risk assessment must reflect the scale of the environment being assessed. Part II focuses on the machinery to identify the relevant scaling relation and thereafter articulate its operational implications.

Part II
Stochastic Security Risk Management

Chapter 5
Security Risk Management Statistics

5.1 Introduction

Modern IT environments can contain thousands if not hundreds of thousands of individual elements. Each of these elements likely contains risk factors for information compromise. These environments are connected to other environments through intermediary networks and are continuously changing as users log on, log off, send email, connect to remote networks and access internet sites.

Information is collected, stored, processed and shared via technologies located on and off-premise. As a result, risk factors for information compromise proliferate throughout this vast ecosystem that supports network users who work with information-related processes, workflows and technologies.

Larger swaths of an IT environment contain more elements and therefore the potential for information compromise is somehow proportional to the accretion of risk factors. The details regarding proportionality are risk-relevant. Specifically, how has the potential for information compromise changed in IT environments containing 100 versus 1000 risk factors? If the potential for information compromise is a function of the number of risk factors, clearly the scale of the assessment matters.

However, an assumption of scale-dependence might suggest security risk management is merely an exercise in counting. One practical problem that arises from this view is risk factors for information compromise affecting IT environments can be difficult to identify let alone count with precision. In addition, focusing exclusively on the number of risk factors ignores the effect of security risk management.

In that vein, cybersecurity risk management is never perfect. Each application of a security control to a risk factor is inherently uncertain with respect to risk management effectiveness. Therefore, an overall estimate of risk management uncertainty for a given IT environment must reflect the uncertainties of all risk factors in aggregate. Furthermore, such an estimate would presumably be relevant to any scaling relation that characterizes cybersecurity risk on an enterprise scale.

A statistical approach to characterizing security risk management seems necessary in order to identify this scaling relation. Therefore, risk factor variability must be

C. S. Young, *Cybercomplexity*, Advanced Sciences and Technologies for Security Applications, https://doi.org/10.1007/978-3-031-06994-9_5

made irrelevant, where risk factors and security controls must be "homogenized" to facilitate statistical analyses. To that end, the notion of IT environment states is introduced next pursuant to developing the necessary statistical machinery.

5.2 IT Environment States

IT environments are ecosystems of network users, information management processes and information technologies that enable network users to access electronic information. The possible configurations of a modern IT network are limitless, where the configuration details have implications to both information sharing and the potential for information compromise.

Each element of an IT environment contributes to the information management mission. Equally, each element has risk factors that increase one or more components of cybersecurity risk. Tools designed to provide visibility, restrict on-line activity and validate access privileges must be applied across this ecosystem that possesses a dynamic security risk profile.

Attempting to identify and address every risk factor would be exhausting and ultimately ineffective. Yet somehow all risk factors must be accounted for in an enterprise model of cybersecurity risk. The problem is to achieve a broad perspective without being overwhelmed by details.

We could naively count risk factors to arrive at a crude, macroscopic metric for cybersecurity risk. This approach is unsatisfying on several levels. Most notably it ignores the effect of security risk management. A second approximation would include security controls in our count. The difference in the number of risk factors and security controls would be indicative of cybersecurity risk on an enterprise scale.

The second order approach is only slightly more satisfying than the first order approximation. Security risk management is imperfect, and the latest approach neglects to account for this imperfection. The imperfection might be trivial when considering a small number of risk factors, but the effect on cybersecurity risk will be amplified as the assessment scale is widened. We need to somehow account for the lack perfection and zero in on the scale-dependence.

The second order approximation also assumes perfect alignment between security controls and risk factors. For example, if there are ten risk factors and nine security controls, all nine security controls are assumed to be matched to the corresponding risk factors. The gap between the two groups is indicative of the magnitude of cybersecurity risk within that IT environment.

Of course, the nine security controls could be misapplied and/or inadequate for the job. If there are indeed only ten risk factors we could assess each one and determine its status. What if there are 1000 or 10,000 risk factors that are changing with time? Risk assessments on that scale become intractable.

To account for security risk management imperfection on an enterprise scale, a probability distribution could be used to describe the inherent uncertainty of the cybersecurity risk management process. Specifically, we could assume security risk management is a random variable with two possible outcomes: a risk factor managed by a security control or a risk factor is not managed by a security control.

Such a process is analogous to a coin toss, where the "heads" and "tails" outcomes are now replaced by "managed" and "unmanaged" risk factors. Any stochastic process with exactly two outcomes is termed a *binary* stochastic process, and the outcomes of that process are characterized by a binomial distribution. For now the probability of a managed risk factor is assumed to be equal to the probability of an unmanaged risk factor.

Repeated application of this risk management process yields a spectrum of unique combinations of managed and unmanaged risk factors. Each of these combinations constitutes an IT environment *state*. The key point is all risk factors and security controls are equivalent and any variations in individual risk factors have become irrelevant. In addition, the probabilities associated with the security risk management process reflect the magnitude of risk management uncertainty across the IT environment.

The challenge, which will be addressed in Chap. 6 via the introduction of information entropy, is to incorporate the magnitude of security risk management uncertainty into this binary security risk management process. The ultimate objective is to account for the effect of risk management uncertainty across a given IT environment pursuant to identifying a scaling relation for enterprise cybersecurity risk.

Predicting the distribution of the resulting states becomes a straightforward exercise if a binary probability distribution is used to approximate security risk management. We leverage the familiar coin toss to illustrate the process mechanics. Assume three fair coins are tossed (or equivalently each coin is tossed three times) and there are two possible outcomes per toss, i.e., heads or tails. The coins are fair, which implies the probability of a heads outcome equals the probability of a tails outcome. Since there are three coins, there are $2^3 = 8$ unique states resulting from the tossing process. Therefore, the probability any particular state results from the coin toss equals 1/8 or 0.125. The distribution of states is shown in Table 5.1.

Merely substituting coins with risk factors yields the identical result only with different labels noting the risk factors are also indistinguishable. In other words, each

Table 5.1 States of a coin toss using three fair coins

State	Probability	Coin #1	Coin #2	Coin #3
State 1	0.125	H	H	H
State 2	0.125	H	H	T
State 3	0.125	H	T	T
State 4	0.125	H	T	H
State 5	0.125	T	T	H
State 6	0.125	T	H	T
State 7	0.125	T	H	T
State 8	0.125	T	T	T

of the IT environment states resulting from risk management consists of a unique combination of managed and unmanaged risk factors that occurs with a 1/8th or 0.125 probability.

In general if the probabilities of the two risk management outcomes are equal and the IT environment contains N risk factors, 2^N states will result from the security risk management process. Each of these states is equally probable. Therefore, the likelihood the environment exists in any particular state is 2^{-N}. That probability increases or decreases exponentially with increasing or decreasing values of N. As a reminder, each state consists of a unique combination of managed and unmanaged risk factors.

Progress has been made in achieving the immediate goal of homogenizing IT environments. However, a more nuanced approach is required if this model is to approximate reality. Namely, a model of cybersecurity risk management must somehow incorporate the effects of security risk management within this probabilistic framework. This effort will be accomplished by quantifying the uncertainty of probability distributions using basic concepts drawn from information theory.

5.3 Information Content and Message Source Uncertainty

Traditional views of information emphasize content or meaning. Yet suppose the interest is in quantifying information? Specifically, what if the quest is to determine the limit on information transmission within a noisy channel? The latter is the problem Claude Shannon tackled in his seminal work, *A Mathematical Theory of Communication.*[1] Note this work was later published as a book with a similar title but authored by Shannon and Weaver. Both works are referenced throughout this text.

We will go out on a limb and claim information is impossible to quantify if its meaning is relevant. For example, does the message, "I saw a big cat" contain more or less information than "The matrix eigenvalues are 3 and 4?" If the meaning of each phrase matters, the answer to that question is both subjective and contextual.

Furthermore, what if the first message were communicated in Chinese but the message recipients cannot speak a word of that language? Clearly the message has meaning to the billions of Chinese speakers but will be gibberish to anyone else. In other words, the meaning of the information contained in a message also depends on the message recipients.

In many circles the statement about eigenvalues would be utterly meaningless no matter what language is used to transmit the message. Yet to a mathematician or physicist this message might reveal the answer to a longstanding problem.

The sentence about the big cat is not likely to inspire great thoughts. Yet to a zoologist studying feline behavior the sentence might suggest she was on the trail of an interesting research subject. That said, a three year-old could easily utter the

[1] C. Shannon, *A Mathematical Theory of Communication*, Bell System Technical Journal, 1948.

same phrase, which implies it lacks any semblance of profundity with apologies to any three year-old readers.

The lesson is the traditional view of information is unsuitable for quantifying information transmission. If information meaning is irrelevant, the aforementioned limit will likely relate to quantities such as the number and rate of transmitted symbols. Therefore, if information is recast as symbols that are generated by a stochastic process or message source, the content automatically becomes irrelevant. Specifically, if information can be related to the *diversity* of the stochastic process/information source, information quantification is straightforward as demonstrated below.

Consider an email consisting of a single letter repeated ad infinitum. It is clear that nothing useful is being conveyed except perhaps the lack of creativity of the author. Note only one symbol is required to communicate the entire email. Furthermore, the resulting message is completely predictable since the probability of a given symbol is always a certainty. In other words, the source of symbols used to generate each message has zero diversity.

Suppose we characterize messages strictly in terms of their predictability. In the case of the single-letter email there is complete certainty and hence equal predictability regarding the identity of each symbol within the message. For example, if the repeated symbol is the letter "C," the probability any symbol selected at random from the message is "C" equals one. Any message derived from a message source where the probability of a particular symbol is one is guaranteed to be completely predictable.

Said another way, any message derived from a source with zero diversity is completely predictable. In contrast, a message derived from a source comprised of (say) 26 symbols would be significantly (and quantifiably) less predictable. If each symbol in the source appears with equal probability, the uncertainty associated with this message source is a maximum.

This condition implies the diversity of the message source is also a maximum. We can see this point more clearly if we think of diversity as the uncertainty density. If the probabilities associated with the outcomes are equal, no particular outcome is favored. Therefore, there is no way to know which outcome will appear next, i.e., there is maximum outcome uncertainty.

Since the sum of the probabilities must equal one by definition, the probabilities *per outcome* are equal, and therefore uncertainty is maximally distributed across the distribution. This condition implies the diversity of the outcome probabilities is a maximum. We repeat for emphasis, the information in this context relates to the diversity of the message source, which determines the predictability of the transmitted messages.

Thankfully, the source of symbols used to compose emails in English actually consists of 26 letters plus a space, i.e., 27 symbols in total. Of course, messages written in English do not appear with equal probability. Since English words and sentences are subject to rules of grammar and spelling, these constraints impose a structure that affects the predictability of specific letters and combinations of letters.

If the diversity of a stochastic process/message source can be quantified, the predictability of messages so derived is also quantifiable. Note a stochastic process can be considered a message source. Recalling Shannon's original objective of determining the limits of signal transmission in the presence of noise, it turns out message source diversity is key to determining this limit, where the latter is known as the *channel capacity*. Furthermore, the greater the diversity of the message source the more symbols are required to transmit messages derived from that source.

For example, contrast the redundant email with one comprised of randomly generated letters derived from the English alphabet. If each letter appears with equal probability, the probability any given letter will appear next equals 1/27.[2] The email message conveys plenty of information in the information-theoretic sense but has no meaning in the traditional sense.

Any language where each symbol in the source alphabet appears with equal probability would not be much use for communicating meaning. Of course, anyone reading this book already appreciates the fact that letters do not occur with equal probability in English. This statement presumably applies to all languages since the purpose of language is to communicate information in the traditional sense, i.e., relate content.

The frequency of letter occurrences can be leveraged to optimize information transmission. Word frequencies for letters and combinations of letters in English have been calculated.[3] For example, the combination of t and h to form "th" occurs with relatively high frequency. In contrast, the combination of q and x to form "qx" occurs with much lower frequency.

The redundancy of the symbols in a source alphabet affects the information-per-symbol required to transmit messages derived from that source. Specifically, the more symbol redundancy the less information per symbol is required to transmit messages so derived. Recall the email derived from a single-letter message source. In that case the required information per symbol is a minimum since the probability of any particular symbol is one. Again, the message source has zero diversity.

We now turn our attention to cybersecurity and applying Shannon's view of information to security risk management. What if a binary stochastic risk management process is the information source and security risk management is analogous to a *biased* coin toss. In other words, the two risk management process outcome probabilities might actually differ from 0.5. The effect of risk management is to bias the security risk management process in favor of either managed or unmanaged risk

---

[2] This statement assumes the probability of letter appearance in English is equal for all letters. If this condition were true, the entropy would be 4.76 bits/symbol (27 symbols consisting of 26 letters plus a space). In reality the English language is redundant, therefore the information entropy of the English language has been calculated to be 4.03 bits/symbol (H. Moradi, J. W. Grzymala-Busse, J. A. Roberts, *Information Sciences, An International Journal* 104 (1998), 31–47).

[3] Zipf's law states that given a large sample of words, the frequency of any word is inversely proportional to its rank in the frequency table. So word number n has a frequency proportional to $1/n$. The simplest case of Zipf's law is a simple power law, 1/f. Thus the most frequent word will occur about twice as often as the second most frequent word, three times as often as the third most frequent word, etc.

factors in exact analogy with an asymmetric coin favored to land on either heads or tails.

The immediate challenge is to quantify the diversity of the message source. Fortunately, Shannon provides a way forward. His derivation of the theoretical limit of channel capacity is based on message source *entropy*, which quantifies the uncertainty/diversity of a probability distribution in terms of the logarithm of the probabilities.

Recall the diversity of the message source determines the unpredictability of messages so derived. IT environment states are the equivalent of messages in this context, and the stochastic risk management process determines the unpredictability of the resulting states. Jumping ahead, we will see that the magnitude of unpredictability equates to complexity in this context.

The immediate task is to quantitatively relate security risk management to the statistics of the risk-managed states. That relationship is achieved by assuming a probabilistic model for cybersecurity risk management and by quantifying the relevant probability distribution via Shannon's message source entropy.

Chapter 6
Information Entropy

Information: the negative reciprocal value of probability.
Claude Shannon

6.1 Introduction

Each element in an IT environment contributes to the magnitude of cybersecurity risk. However, their aggregate effect can be difficult to ascertain. Furthermore, the variability of IT environments coupled with the sheer number of elements preclude assessments of every element and aggregating the individual results.

Therefore, a statistical approach is perhaps the only recourse if the objective is to develop a *macroscopic* view of cybersecurity risk. The key to achieving that objective is assuming cybersecurity risk management is a stochastic process. As mentioned at the conclusion of Chap. 5, there needs to be a quantitative link between the diversity of a stochastic risk management process/information source and the resulting states consisting of managed and unmanaged risk factors in order to develop the requisite statistical profile.

That link is provided by the information or Shannon entropy.

The concept of entropy has been appropriated from information theory, where a stochastic security risk management process is considered a message source. Once such a model has been adopted, the mathematics used to characterize complexity is straightforward, where the calculations are identical to those describing the outcomes of a coin toss.

The operational implications of such an assumption represent a potential source of angst. Declaring security risk management to be a binary stochastic process may be mathematically kosher, but ultimately its connection to reality must be reconciled in order to generalize the results to more deterministic scenarios.

Perhaps contrary to appearances, leveraging information entropy attempts to add a semblance of realism by introducing security risk management into an otherwise

C. S. Young, *Cybercomplexity*, Advanced Sciences and Technologies for Security Applications, https://doi.org/10.1007/978-3-031-06994-9_6

random process. The entropy reflects the degree to which security risk management decision-making deviates from complete randomness, i.e., a fair coin toss. However, there is likely still some explaining to do before launching into the details.

The use of entropy has been mathematically justified based on its similarity to a coin toss, a topic that will be covered in the discussion of ergodicity. It has been operationally justified based on expedience and the reasonableness of model results.

That said, a stochastic model of security risk management is not a panacea for the shortcomings that plague all or nearly all cybersecurity risk models. Namely, the value of entropy and hence the magnitude of complexity/enterprise cybersecurity risk cannot be calibrated. In other words, it is not possible to precisely correlate the value of entropy with the magnitude of complexity. The last sentence implies it is equally impossible to adjust security controls so that complexity is reduced to some quantifiable threshold. There can be no complexity meter that can be precisely adjusted by dialing up or dialing down security controls.

The stochastic approach, and the use of entropy in particular, is a means of developing general insights into complexity within IT environments and in identifying the types of security controls required to address it. However, even these limited results represent incremental advances, especially with respect to providing operational guidance on countermeasures.

Finally, the discussion on entropy is somewhat technical although it is still elementary by a mathematician's standards. A more sophisticated treatment is not needed nor is it particularly relevant. It is merely necessary to understand what information entropy is, why it is relevant to IT environments, and how it is applicable in this context.

Unfortunately or not, the above means some mathematics is required. To facilitate understanding worked examples are included in the discussion. However, before that discussion begins we must understand the types of processes to which entropy can be applied, and why security risk management as defined herein fits the bill.

6.2 Ergodicity

A security risk management process/message source cannot be just any stochastic process in order for information theory to be applicable. It must be *ergodic*, where the criteria for ergodicity are as follows: (1) the process is stationary and (2) the ensemble average of process outcomes equals its time average. These concepts will be explained in this section.

The ergodicity of the security risk management process used to model IT environment complexity will not be proven rigorously. The "proof" is by analogy, where the process is equivalent to a coin toss, which is indeed ergodic. However, an explanation of ergodicity is still required.

An information source producing a sequence of outcomes is deemed stationary if the average value of the sequence is independent of where the sequence begins. For example, the sequence ABABABABABABABAB is stationary whereas ABAAB-BAAABBBAABA is not. In the former sequence as with all stationary processes, past is prologue.

If a sub-sequence of symbols from the former sequence were selected at random, it too would be stationary. Clearly time and translational-invariance are both properties of a stationary source.

Next consider a finite sequence of characters where the probability of a character is dependent on the character that preceded it, and assume these probabilities vary by character. In general, the initial values of the source will affect the statistics of the sequence of characters for some finite distance from the beginning character.

A mathematical work-around is to not just consider one sequence of characters produced by the source. Rather, if the source is initiated an infinite number of times it will produce an infinite number of sequences of characters. Such a collection is called an *ensemble* of sequences.

Next line up these sequences in parallel. If some fraction of the sequences begin with specific characters, e.g., 10% of the sequences begin with character type A, and 15% begin with character type B, etc., and the average is calculated by position across the ensemble, the average frequency by character type will be the same no matter where in the sequence the count begins. Figure 6.1 illustrates the concept of an ensemble average.

In other words, the probability by character type is independent of position. This condition is what is meant by stationary. Any statistic obtained by averaging over the

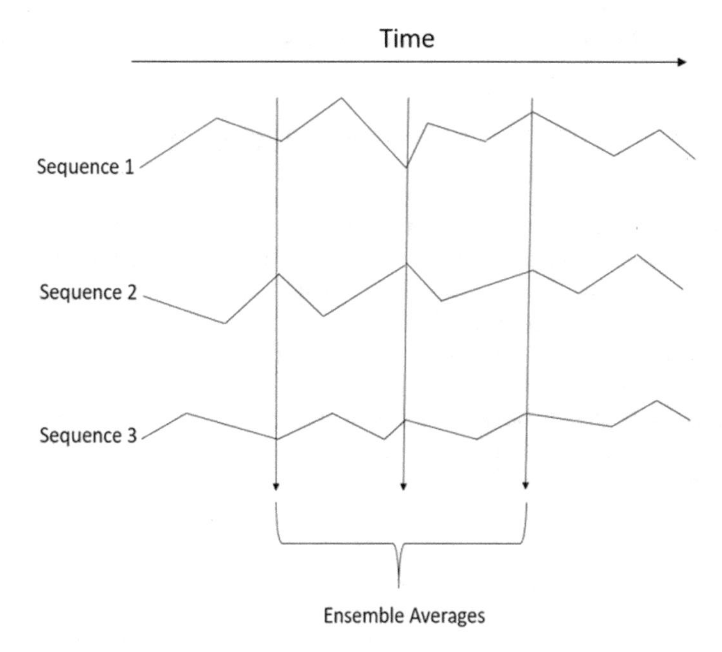

Fig. 6.1 Ensemble average

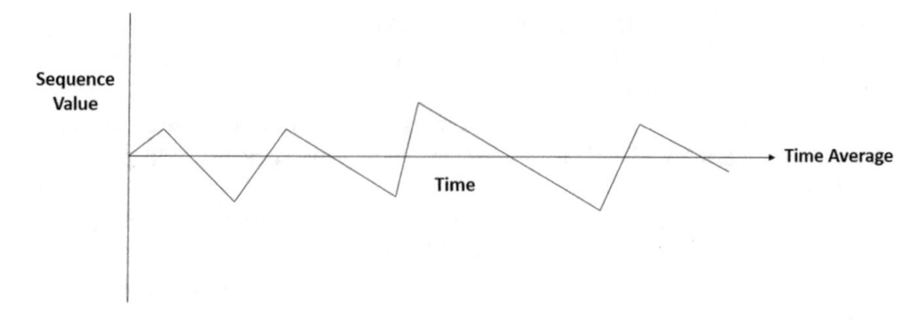

Fig. 6.2 Time average

ensemble does not depend on the distance from the beginning of the average, and is therefore valid at any position or time in the continua.

There is another way of calculating averages over sequences of values. Namely, concatenate all sequences and average these over time, i.e., horizontally across a sequence. Figure 6.2 illustrates a time average over a sequence of process outcomes.

A process is called *ergodic* if the time average and ensemble averages are equal and the process is stationary. Ergodicity becomes especially important if there is only one sample function or sequence resulting from a stochastic process rather than the entire ensemble, i.e., a collection of sequences. A single sample function will often provide little information about the statistics of a process.

A mathematical definition of ergodic for some function A is as follows, where s is the function's value at some point in a sequence of values and t equals time,[1]

$$\lim_{\Delta t \to \infty} \frac{1}{\Delta t} \int_{t}^{t+\Delta t} A(s)ds = \lim_{N \to \infty} \frac{1}{N} \sum_{i}^{N} A_i(t)$$

If a process is ergodic, all statistical information can be derived from just one sample. One sample is representative of the entire process, and selecting one sequence in Fig. 6.1 and taking the time average will yield the average of all other sequences. Clearly, a process must be stationary for this procedure to be valid. Furthermore, ergodicity implies stationarity but not the other way around.

Examples might provide clarity.[2] Consider a collection of batteries of various types. Suppose a specific battery is selected at random and its voltage is measured. This voltage v(t) represents one particular voltage selected from a subset of constant, and possibly different, battery voltages.

[1] A Simple Explanation of Ergodicity in Finance, https://medium.com/@mhegdekatte/a-simple-explanation-of-ergodicity-in-finance-part-i-7b6892433645.

[2] https://www.nii.ac.jp/qis/first-quantum/e/forStudents/lecture/pdf/noise/chapter1.pdf.

This voltage measurement process is stationary but not ergodic, where the time average only corresponds to the voltage of the particular battery selected. The statistical average will likely be another voltage, depending on the voltages of the other batteries. Therefore, the process is not ergodic in the mean.[3]

Another more dramatic example of a non-ergodic process is as follows.[4] Imagine a game of Russian roulette using a revolver with six chambers, and a bullet is loaded into one chamber. If you choose to play the game and live you receive one dollar, which hardly seems worth it but that is beside the point.

Suppose six people play the game and each person pulls the trigger once during the game. Barring misfires, at the end of the game 83% (5 players) will be one dollar richer, and 17% (1 player) will transition to the next world. Now instead of six people playing the game once, suppose one person plays the game six times. Two things are certain: the naïve player will not be receiving his or her one-dollar reward and Russian roulette is not an ergodic process.

Another way of thinking about ergodicity is in terms of individual versus collective behavior. For example, a process where one person tosses n coins is statistically equivalent to n individuals tossing a single coin. A coin toss is an example of an ergodic process.

In summary, an ergodic process is a process where the behavior over time of a random process exhibits the same behavior as multiple identical processes at a single point in time. Ergodicity is a prerequisite for characterizing stochastic processes using information entropy.

Further discussions of ergodicity are beyond the scope of this book. Since the security risk management process adopted herein is exactly analogous to a coin toss, an assumption of ergodicity seems safe.

6.3 Introduction to Information Entropy

The prospect of applying information theory to security risk management is cause for deliberation. The physicist John R. Pierce describes the situation beautifully.[5]

> This difference between the exactly ergodic source of the mathematical theory of communication and the approximately ergodic message source of the real world should be kept in mind. We must exercise a reasonable caution in applying the conclusions of the mathematical theory of communication to actual problems. We are used to this in other fields.

Pierce continues by saying,

> Whatever caution we invoke, the fact that we have used a random, probabilistic, stochastic process as a model of man in his role of a message source raises philosophical questions.

---

[3] Ergodicity can apply to either the mean or correlation.

[4] Op. cit. https://medium.com/@mhegdekatte/a-simple-explanation-of-ergodicity-in-finance-part-i-7b6892433645.

[5] Pierce, John R. An Introduction to Information Theory, Symbols, Signals and Noise, Dover Publications, Second Edition, 1980.

> Does this mean that we imply that man acts at random? There is no such implication. Perhaps if we knew enough about a man, his environment, and his history, we could always predict just what word he would write or speak next.
>
> In communication theory, however, we assume that our only knowledge of the message source is obtained either from the messages that the source produces or perhaps from some less-than-complete study of man himself. On the basis of information so obtained, we can derive certain statistical data which, as we have seen, help to narrow the probability as to what the next word or letter of a message will be. There remains an element of uncertainty.
>
> For us who have incomplete knowledge of it, the message source behaves as if certain choices were made at random, insofar as we cannot predict what the choices will be. If we could predict them, we should incorporate the knowledge which enables us to make the predictions into our statistics of the source. If we had more knowledge, however, we might see that the choices which we cannot predict are not really random, in that they are (on the basis of knowledge that we do not have) predictable.

Pierce's warning about applying the mathematical theory of communication to other problems is sobering. In this instance the assumption of a stochastic risk management process is driven mostly by expedience with little or no expectation of precision. Nevertheless, the inherent variability of IT environments suggests uncertainty is always present.

Therefore, a stochastic process is leveraged to model cybersecurity risk management, and information entropy can be used to quantify the uncertainty of the relevant probability distribution. But first we need to understand information entropy and its relationship to probability distributions, uncertainty and diversity.

Probabilities are actually a reflection of ignorance. An example seems appropriate to help justify such a statement.

Consider a set of N states and a subset of M states, where each state in the subset has equal probability. All remaining states, i.e., N – M states, have zero probability. Figure 6.3 illustrates the situation.

Each state in the distribution of M states has probability 1/M. This probability defines the magnitude of certainty, or conversely, the uncertainty in selecting a given state at random from M.

Next consider a different distribution, where Q states are a subset of N states, and each state in Q has equal probability. The N – Q states have zero probability. Figure 6.4 illustrates this situation.

The probability of identifying a given state from the subset of Q states is higher in Fig. 6.4, i.e., 1/Q, since Q is smaller than M. Simply put, in the distribution of Fig. 6.4 there are fewer states from which to choose. Therefore, the uncertainty, i.e., *ignorance*, regarding the subset of Q states has decreased relative to the M states in Fig. 6.3.

In other words, the magnitude of *uncertainty* is *directly* proportional to the number of states in the distribution possessing a non-zero probability. Conversely, the magnitude of *certainty* is *inversely* proportional to the number of states in the subset.

One could just as easily define this uncertainty in terms of the logarithm of the probability rather than the probability itself. In other words, define the uncertainty associated with a random variable with probability p to be equal to—log p. The minus

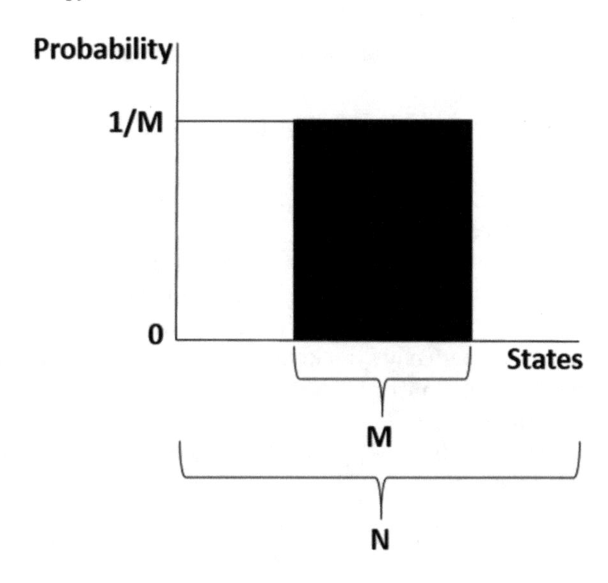

Fig. 6.3 M states with equal probability and N total states

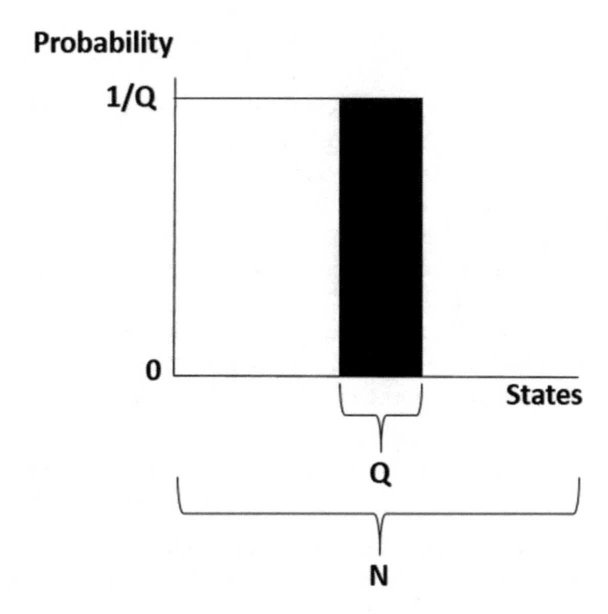

Fig. 6.4 Distribution of Q states with equal probability and N total states

sign merely ensures uncertainty has a positive value since p is always less than one by definition. This expression is defined as the *information* or *Shannon entropy* of a probability distribution, or just *entropy* for short.

Of course, the aforementioned examples were chosen because of their simplicity. What if a probability distribution consisted of different probabilities? One might specify a weighted average of these probabilities in estimating uncertainty. Differences in probabilities lead to a more general definition of entropy as follows.

The entropy (H) of a probability distribution with n elements is given by,

$$\overset{n}{\underset{i}{H}} = -\sum p_i \log_2(p_i) \tag{6.1}$$

The subscript indicates the logarithm is base 2. Readers might recognize (6.1) as the average of log (p_i).

The now-familiar example of a coin toss illustrates information entropy in action. If an individual tosses a single coin in the air, the probability of a heads outcome equals the probability of a tails outcome to a good approximation. In other words, if the coin is perfectly symmetric about the principal axis of rotation, there is exactly a fifty percent chance of either a heads or tails outcome. What is the entropy associated with this coin toss?

Applying Eq. (6.1) yields the following expression, where p(heads) = p(h) and p(tails) = p(t),

$$\begin{aligned} H &= -\left[p(h)\log_2 p(t) + p(t)\log_2 p(t)\right] \\ &= -[0.5(-1) + 0.5(-1)] \\ &= 1 \text{ bit per toss} \end{aligned}$$

Since the logarithm is in base 2 the units of H are information "bits." Therefore, one information bit is required to specify the outcome of a single coin toss. As noted previously, a stochastic process can be considered an information or message source, and the greater the diversity of the source the more entropy per outcome is required to specify a given outcome. As a prelude of things to come, consider the impact on H if a binary stochastic process has unequal probabilities. In other words, what if p(h) does not equal p(t)?

In 1948, Shannon recognized the concept of information entropy was actually not new, wherein he realized it is related but not equivalent to thermodynamic entropy. Shannon comments on this issue in his classic paper.[6]

> The form of H will be recognized as that of entropy as defined in certain formulations of statistical mechanics where p_i is the probability of a system being in cell i of its phase space. H is then, for example, the H in Boltzmann's famous H theorem.

John Pierce provides an elegant explanation of how information entropy and thermodynamic entropy are related.[7] Both types of entropy convey uncertainty. Information entropy is a measure of the uncertainty with respect to what message, among many possible messages, a message source will produce for any given message transmission.

---

[6] C. Shannon, W. Weaver, A Mathematical Theory of Communication, Bell System Technical Journal, 1948.

[7] J. Pierce, op. cit.

Thermodynamic entropy represents the uncertainty regarding the state of a physical system, i.e., k ln (Ω), where k is a constant of proportionality known as Boltzmann's constant and Ω is the number of possible states of the system. Physical states consist of atoms and molecules with various positions and momenta.

The two types of entropy are related in that the information resulting in a reduction of the thermodynamic entropy of a system is not cost-free. The cost is proportional to the information entropy of the message source that produces the information regarding the state of a system.

We need to explore the precise connection between the diversity of a stochastic process/message source and the unpredictability of the process results. The nexus is supplied by the stochastic process/message source entropy, where we will ultimately describe complexity in terms of the unpredictability of the probable states that result from a stochastic security risk management process/message source.

Recall from Chap. 5 the probabilities of the heads and tails outcomes of a coin toss combined with the number of tosses determined the number of resulting states, where each state consisted of a unique mixture of heads and tails outcomes. For example, if 1000 fair coins are tossed once (or equivalently one coin is tossed 1000 times), 2^{1000} probable states will result from this process. Note this figure represents the *maximum* number of probable states when 1000 coins are tossed.

The probability the process will yield any *particular* state from the spectrum of probable states equals 2^{-1000}. This result assumes the coin (or coins) is fair, i.e., $p(h) = p(t) = 0.5$. Therefore, there is maximum process uncertainty, or equivalently, maximum diversity. However, if the coin is biased, a lesser number of probable states will result. Therefore, the unpredictability of the resulting states will also decrease.

Although coin fairness is typically taken for granted, recognize it is merely an assumption the next time significant stakes are on the line. A more precise definition of fairness is the probability distribution of the two binary stochastic process outcomes is maximally diverse.

Importantly, information entropy quantifies the inherent diversity of any probability distribution.

The information entropy is a maximum when $p(h) = p(t) = 0.5$ in a binary stochastic process. Since the information entropy reflects the uncertainty of a probability distribution, the information entropy of a coin toss or any binary stochastic process is also a maximum when the process is fair, i.e., the two probabilities are equal.

If the coin is fair it is impossible to predict the outcome of a single coin toss, which is presumably the point of tossing a coin as a method of decision making. However, it is possible to determine the likely distribution of a large number of coin tosses, which is arguably an objective of statistics.

As noted above, if a coin were asymmetric about a principal axis of rotation it would introduce bias into the coin toss process. Clearly if a biased coin were tossed a thousand times, the resulting states would be affected by this asymmetry, where some states would be more probable than others depending on whether the coin was biased in favor or heads or tails.

Now suppose the probability of one of the process outcomes is a certainty. In other words, the probability of one of the two outcomes is one. The probability of the complementary outcome must be zero since the sum of the two probabilities must equal one by definition. What would the distribution of the resulting states look like following one thousand process events?

If the probability of one of the two outcomes were a certainty, the resulting states would consist entirely of that particular outcome. Therefore, the probability any particular state consists entirely of that outcome would be one, and the probability a state consists of the other outcome would be zero. Moreover, the unpredictability of the resulting states is a minimum since all states are identical. Put another way, there is only one probable state, therefore the resulting states are completely predictable.

Maximum and minimum security risk management diversity correspond to entropy values of one and zero, respectively. It is certainly possible the probability distribution associated with the risk management process falls somewhere between these extremes. In other words, each of the two outcome probabilities could be greater than zero and less than one, which means the entropy also lies somewhere between zero and one.

Figure 6.5 reveals the states resulting from biased, partially biased and unbiased binary stochastic processes.

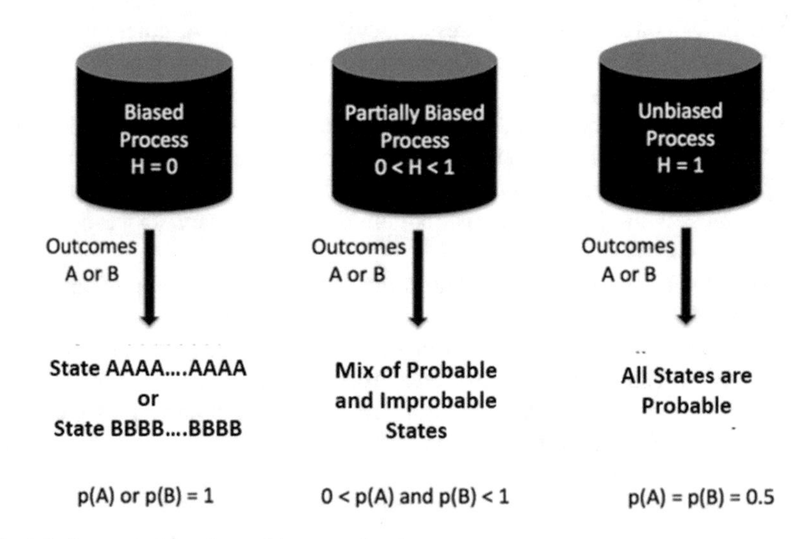

Fig. 6.5 States resulting from a binary stochastic process

6.4 Applying Information Entropy

One might be inspired to ask why the probabilities themselves are not used to characterize uncertainty/diversity rather than the logarithm of the probability. Claude Shannon himself supplies the answer[8]:

1. *It is practically more useful. Parameters of engineering importance such as time, bandwidth, number of relays,* etc., *tend to vary linearly with the logarithm of the number of possibilities. For example, adding one relay to a group doubles the number of possible states of the relays. It adds 1 to the base 2 logarithm of this number. Doubling the time roughly squares the number of possible messages, or doubles the logarithm,* etc.
2. *It is nearer to our intuitive feeling as to the proper measure. This is closely related to (1) since we intuitively measure entities by linear comparison with common standards. One feels, for example, that two punched cards should have twice the capacity of one for information storage, and two identical channels twice the capacity of one for transmitting information.*
3. *It is mathematically more suitable. Many of the limiting operations are simple in terms of the logarithm but would require clumsy restatement in terms of the number of possibilities.*

One might also ask why measure information entropy and not some other quantity? Again Shannon provides the answer, where any quantity resembling H must satisfy the following three key conditions:

1. It should be continuous in probabilities.
2. It should increase as the number of possible outcomes increases. There is more uncertainty with more possible events.
3. If a choice is decomposed into successive choices, the original value should represent the weighted sum of individual values.

Shannon continues by saying,

> This theorem, and the assumptions required for its proof, are in no way necessary for the present theory. It is given chiefly to lend a certain plausibility to some of our later definitions. The real justification of these definitions, however, will reside in their implications.[9]

The information entropy enables comparisons of security risk management strategies that are modeled as stochastic processes. In the simplest of examples, the information entropy can be used to quantify a "security by obscurity" strategy, and thereby compare it with competing strategies.

Assume an IT environment has a total of n identical subnets. The probability an attacker targets a specific subnet via random selection is of course, $1/n = p_1$. The information entropy of this message source is given by (6.1),

[8] C. Shannon, W. Weaver, *The Mathematical Theory of Communication*, University of Illinois Press, 1949.

[9] Ibid.

$$
\begin{aligned}
H_1 &= -n \times p_1 \log(p_1) \\
&= -\log(1/n)
\end{aligned}
$$

Suppose each subnet contains an identical number of information assets. The probability of selecting the correct asset within a given subnet containing m assets is given by p_2, which as expected equals 1/m. Therefore, the entropy associated with asset selection within a subnet equals,

$$
\begin{aligned}
H_2 &= -m \times p_2 \log(p_2) \\
&= -\log(1/m)
\end{aligned}
$$

However, the probability of randomly selecting a particular asset from within a randomly-selected subnet equals the product of p_1 and p_2, i.e., $1/n \times 1/m = 1/(nm)$. The probability of guessing the correct asset decreases exponentially with the product of the subnets and the assets-per-subnet. Since p_1 and p_2 are independent, the entropy associated with selecting both the correct subnet and the correct asset within the correct subnet, i.e., the joint entropy (H_t) equals,

$$
\begin{aligned}
H_t &= H(p_1 \times p_2) \\
&= H(p_1) + H(p_2) \\
&= -\log(1/n) - \log(1/m)
\end{aligned}
$$

Therefore, since the objective is to maximize uncertainty for the attacker, the strategy would be to increase H_t, which implies increasing n and/or m. The optimum defensive strategy to counter an external threat is to maximize unpredictability in selecting an information asset. Such a strategy is antipodal to the internal risk management strategy, which favors increasing IT environment uniformity as we will discuss in Chaps. 10 and 12.

Note that if assets are *uniformly* distributed among the subnets, the protection afforded by subnets against random target selection disappears. To see this effect, assume there is a total of six information assets and no subnets. One asset contains the keys to the information kingdom and is therefore the principal target of an attacker. Trivially, the probability the attacker will randomly identify the asset in question is 1/6 or approximately 0.17.

Now divide the network into three subnets, where each subnet contains two information assets. Again, the probability of identifying the desired asset by random selection p(asset) equals the probability of identifying the correct subnet multiplied by the probability of selecting the correct asset within that subnet.

Specifically, $p(\text{asset}) = 0.33 \times 0.5 \sim 0.17$ as before. Therefore, we see a uniform distribution of assets within subnets confers no operational advantage against random target selection. However, the situation changes if there is an uneven distribution of assets as exemplified by the following example.

Divide the IT environment into three subnets such that the first subnet contains three information assets, the second contains two assets, and the third subnet contains one asset. Assume the subnet with three assets is the one containing the asset of interest.

Once again the probability of selecting the targeted asset is given by the joint probability, which is equal to the probability of selecting the correct subnet times the probability of selecting the correct asset within that subnet. In this case the probability equals $0.33 \times 0.33 = 0.11$, which is clearly less than 0.17.

However, what if the targeted asset is the solitary asset? In that case the probability the attacker identifies the targeted asset on the first try is given by $0.33 \times 1 = 0.33$. Clearly asset placement is relevant to a successful security-by-obscurity defensive strategy.

A better-known example of optimizing uncertainty for attackers relates to passwords. Information entropy is traditionally used to quantify password diversity, which is also known as password complexity. Brute force attacks on passwords involve guessing, where the guessing time is a function of the number of possible passwords and the computational power of the machine used to execute the brute force attack.

The uncertainty associated with password guessing directly relates to the number of possible passwords. The concept of information entropy can be invoked to reflect the probability of a correct guess as measured in bits (base 2). More bits correspond to a larger number of possible outcomes and hence less predictability associated with guessing the correct password. Attackers traditionally crack passwords by attempting to brute force cryptologic hash values of passwords. Cryptologic hashing is a function with specific mathematical properties that make it desirable in protecting against password cracking.

One particularly noteworthy property is the computational infeasibility of two distinct inputs hashing to the same value, i.e., a collision. This property is relevant because sufficient infeasibility would prevent an attacker from intentionally creating blocks of data that violate the assumption each block has a unique fingerprint. The probability of a collision becomes vanishingly small when the number of possible hash values (N) is large. N is made large by using hash values consisting of long bit strings.

The formula for the probability of a hash collision, where k equals the number of randomly generated values of integers, i.e., the number of hash values, is given by the following expression,[10]

$$p(k) = 1 - e^{-k(k-1)/2N} \tag{6.2}$$

Using 32-bit hash values there is a fifty percent chance of a collision when the number of hashes is only 77,163, which illustrates why hashing algorithms such as SHA-256 use 256-bit bit strings.[11]

Figure 6.6 plots expression (6.2) for 32-bit hash values, where $N = 2^{32}$.

[10] https://preshing.com/20110504/hash-collision-probabilities/.

[11] SHA-256 is the hashing algorithm used in Bitcoin.

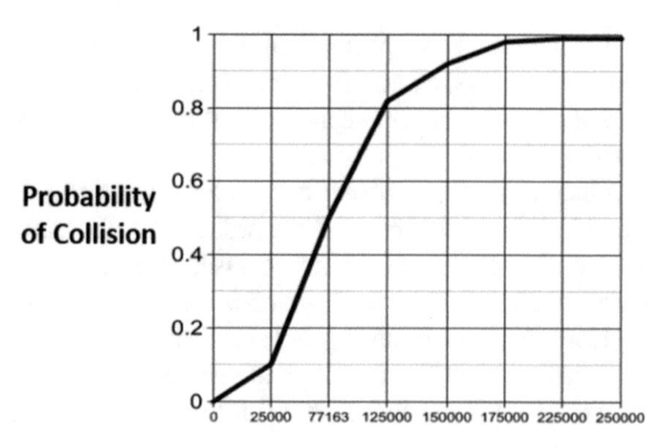

Fig. 6.6 Probability of a collision as a function of the number of 32-bit hashes

Clearly the objective is to increase unpredictability by increasing the information entropy of the password information source. Increasing the time required to create a hash collision means the attacker's computer must be presented with a more computationally infeasible number of possible hash values. It can be shown mathematically that increasing password length has the most significant effect on the number of possibilities and not the diversity of the character set.

Although increasing password length provides protection against potential adversaries, it simultaneously decreases convenience for network users. Convenience and security are characteristically in tension. Because of its widespread impact on network users and its implications to the potential for information compromise, an organization's password policy is an excellent barometer of its security culture. In other words, it is indicative of where the organization exists on the security risk continuum depicted in Fig. 9.1.

What if multiple security functions are modeled as stochastic risk management processes within a given IT environment? How does information entropy scale with the number of processes?

Assume risk management is a random variable X, where p(x) is defined as $p(X = x)$. If there are $x_1, x_2,\ldots x_N$, i.e., N independent processes, where $p(x_1) = p(x_2) = p(x_N) = p(x)$, the total entropy $H(x_1, x_2\ldots x_N)$ is given by $H(x_1) + H(x_2) + \ldots + H(x_N) = NH(p(x))$. In other words, if there are multiple independent stochastic risk management processes within an IT environment, the entropy scales linearly with the number of processes.

In summary, information entropy reflects the diversity of a discrete probability distribution, which characterizes a stochastic process/information source. Security risk management is assumed to be a binary stochastic process/message source, where the two probabilities determine the unpredictability of the resulting IT environment states, and each state consist of a unique combination of managed and unmanaged risk factors. The relationship between information entropy and security risk management will be explored further in the next section.

6.5 Information Entropy and Security Risk Management

Next consider a slightly more complicated example. Assume an information source consists of signals comprised of only ones and zeros. In the modern world such digital message sources are ubiquitous, where analog signals have been digitally encoded to produce music, cell phone transmissions, YouTube videos, etc. Digital encoding is a fancy term for the representation of an analog signal using ones and zeros.

For example, audio speakers inside ear buds respond to various signals generated by an audio amplifier located inside smart phones. The signal voltages are represented by combinations of ones and zeros, which are intended to faithfully reproduce the original audio input.

The more voltages available to encode the analog signal the higher the fidelity since more frequencies can be encoded. If the voltage source consists of eight bits, 2^8 voltage levels can be encoded. If the source consists of sixteen bits the number of levels expands dramatically to 2^{16} voltages.

It is useful to examine specific digital signals in order to obtain further insight into information a la Shannon. For example, what if the audio signal source consisted of eight-bit strings with only ones? In this instance the resulting audio signal would be a single frequency, which would not appeal to most people no matter how mellifluous the tone.

Specifically, an eight-digit message source consisting of only ones would yield only one message: 11111111. Therefore, the probability distribution characterizing this message source would equal 1 for 11111111 and zero otherwise. This distribution is depicted in Fig. 6.7.

Next consider a message source consisting of two eight-digit bit streams, where one of the two streams is 00000000 and the other is 11111111. Each stream has an equal probability of occurrence. In this case the probability of each stream must equal 0.5 and zero otherwise. Figure 6.8 depicts this particular probability distribution.

Finally, consider a message source consisting of zeros and ones where the output is eight digits but each digit can assume a value of either zero or one. In this example

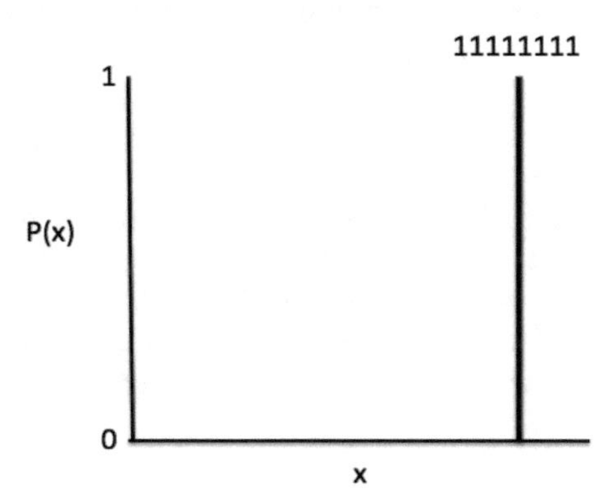

Fig. 6.7 Probability distribution of a single eight-digit outcome

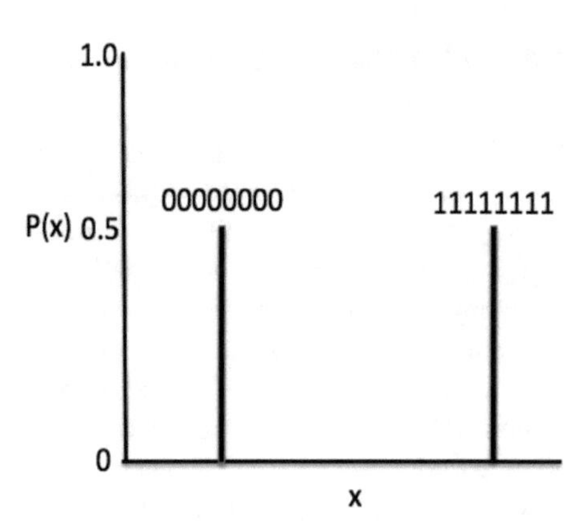

Fig. 6.8 Probability distribution of two eight-digit outcomes

there are $2^8 = 256$ combinations of ones and zeros. Therefore, the probability of a specific combination equals 1/256.

The resulting audio signal has the potential to be considerably more interesting than those derived from the previous message sources due to the difference in the message source diversity. Of course, if the selection process is random it is possible (but unlikely) the same combination of ones and zeros will repeatedly appear. It should be noted the sound resulting from a random selection of highly diverse tones might also not be particularly appealing but that issue is irrelevant to this discussion.

The point is the character of the audio signal relates to the breadth of frequencies that can be encoded via the message source, which in turn is a function of the source diversity. In other words, more diverse sources are capable of encoding more information, where information in this context has a strictly stochastic interpretation.

Claude Shannon proposed a stochastic view of information in his classic 1948 paper cited previously.[12] Furthermore, the information entropy can be used to quantify the diversity of the stochastic process/message source, which Shannon realized could be used to determine the limits on channel capacity.

Equation (6.1) reduces to the following expression if the individual probabilities in a probability distribution are equal since the sum of the probabilities must equal one by definition,

$$H = -\log(p) \tag{6.3}$$

Recall H corresponds to the number of information bits per message source outcome. The information entropy of a binary probability distribution was calculated to be one bit/symbol in Sect. 6.3 of this chapter. In other words, it requires one bit of information to specify the outcome of any binary stochastic process if the probabilities of the two outcomes are equal.

[12] C. Shannon (1948); op. cit.

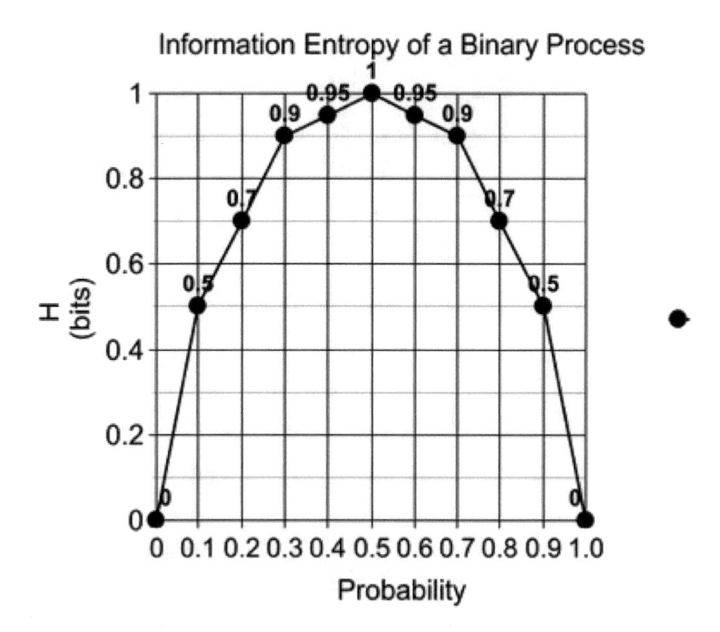

Fig. 6.9 Information entropy of a binary stochastic process

How does the information entropy vary when there is a priori knowledge of the message source outcomes? For example, how does the information entropy of a coin toss change if the coin is biased in favor of either heads or tails?

Suppose p(h) = 0.75 and p(t) = 0.25. The information entropy is again calculated using (6.1),

$$\begin{aligned} H &= -[0.75(-0.415) + 0.25(-2)] \\ &= -[-0.311 - 0.50] \\ &= 0.811 \text{ bits per toss} \end{aligned}$$

In other words, because of the decrease in diversity resulting from the unequal outcome probabilities, only 0.811 information bits are required to specify a message derived from this message source.

Figure 6.9 shows the values of H for *any* binary stochastic process using Eq. 6.1. The graph confirms the entropy has a maximum value when p(outcome 1) = p(outcome 2) = 0.5. H equals zero if either p(outcome 1/outcome 2) = 0 or p(outcome 2/outcome 1) = 1 since in both cases the message source has zero uncertainty.[13] The symmetry of a binary stochastic process has important operational implications to security risk management.

The number of probable states resulting from a coin toss consisting of 1000 fair coins was previously shown to equal 2^{1000}, where 1000 is the number of tosses.

[13] C. Young: *Risk and The Theory of Security Risk Assessment*, Springer Nature, 2019.

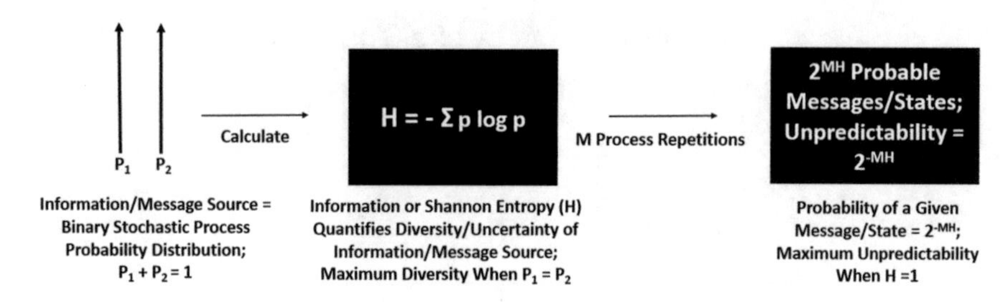

Fig. 6.10 The theoretical foundations of IT environment complexity

The number of probable states resulting from any binary process with entropy H and M independent process events is given by the following,

$$\text{Number of Probable States} = 2^{\text{MH}} \tag{6.4}$$

An example of a process event might be a digital transmitter that transmits one of two possible symbols, e.g., 0 and 1, per event. Therefore, it sends messages consisting of strings of zeros and ones. Pierce provides an example that illustrates the applicability of entropy to quantifying information transmission, and ultimately to the extension of this formalism to information sources within IT environments.[14]

Assume H equals 0.811 bits-per-symbol and 1000 symbols are transmitted by a digital transmitter. According to (6.4) the number of probable states resulting from the transmission process becomes $2^{(1000 \times 0.811)} = 2^{811}$ bits. Note there must also be 2^{1000}–2^{811} *improbable* states, which means each of these states has negligible probability of being transmitted (or received).

Therefore, using 811 binary digits one can write 2^{811} different binary numbers. If one of the binary numbers is assigned to each of the 2^{1000} 1000-digit probable messages, the other improbable messages would go unnumbered.

The key result is one can specify which probable 1000-digit messages an information source produces using only 811 digits. More generally, the number of binary digits or "bits" required to transmit a message is given by the entropy in bits-per-symbol times the number of symbols.

Identical calculations can be applied to IT environments, where security risk management is considered a binary stochastic process/information source. Specifically, the number of probable states resulting from a security risk management becomes a function of the number of IT environment risk factors in combination with IT environment security risk management uncertainty, i.e., the entropy. The resulting figure is a metric for the magnitude of cybersecurity risk on an enterprise scale, which we claim can be used to represent the complexity of an IT environment.

Figure 6.10 summarizes the information-theoretic foundations of IT environment complexity, which will be discussed in more detail in Chaps. 7 and 8.

---

[14] Pierce, J., op. cit.

Part III
Enterprise Cybersecurity Risk

Chapter 7
Complexity and Cybercomplexity

7.1 Introduction

Security control calibration is the Holy Grail of cybersecurity risk management. Although this capability is rare, it does exist. For example, in the WiFi threat scenario described in Chap. 3, security control calibration is possible because signal detection limits are dictated by physics. The Friis formula yields a power law scaling relation equating signal intensity with the inverse-square of distance between a signal receiver and a radiating WiFi access point. Therefore, it is possible to precisely determine the required exclusion zone in order to protect against unauthorized signal detection.

Similarities and differences of physical systems with IT environments can yield insights into cybersecurity risk. In Chap. 1 we compared and contrasted physical systems containing many particles to IT environments possessing numerous risk factors for information compromise. It is instructive to again compare each of these environments with respect to their variability and the implications to measurements in each instance.

A measurement of temperature at one locus of an isolated physical system in equilibrium will yield the same measurement at any other locus according to the Zeroth Law of Thermodynamics. In other words, a local temperature reading of an isolated system in equilibrium is indicative of the temperature of the entire system.

The result of similar "measurements" within an IT environment would likely be vastly different. An assessment of cybersecurity risk confined to a narrow slice of an IT environment would typically not be indicative of the overall potential for information compromise. The particular slice in question might only include a fraction of the total number of risk factors for information compromise. Moreover, the potential for information compromise might change *disproportionately* as additional slices of the environment are included in the assessment. A simple thought experiment reveals the hazards of a myopic view of IT environments.

Consider an IT environment with only one user, one application, one workstation, one router, etc. Such an environment would undoubtedly present a significantly lower risk profile than almost any imaginable IP-based computer network. Next, iteratively

C. S. Young, *Cybercomplexity*, Advanced Sciences and Technologies for Security Applications, https://doi.org/10.1007/978-3-031-06994-9_7

add software, hardware and network users to the original configuration while leaving all other aspects of the environment unchanged.

Each additional element introduces risk factors for information compromise and likely increases security risk management uncertainty. It is clear from this trivial thought experiment that the potential for information compromise is qualitatively greater in IT environments containing a larger number of elements. The fundamental question is not whether the potential for information compromise increases as more slices are added. The more risk-relevant questions are by how much the potential for information compromise increases and what types of security controls should be applied as a result.

In Chap. 6 we learned the information entropy of a binary stochastic process, specified in information bits, quantifies the diversity of the message source. The higher the information entropy the closer the process resembles a fair coin toss, where the maximum value equals one. If the process is applied multiple times, the result is a spectrum of states where each state consists of a unique combination of the two message source outcomes.

At the conclusion of the last chapter we hinted at the possibility of applying this information theoretic formalism to IT environments. The purpose of this chapter is to explore this connection in more detail with the objective of developing a macroscopic view of cybersecurity risk in IT environments. To reiterate, the ultimate objective is to identify a scaling relation for complexity in IT environments and thereby gain insight into the types of security controls required to reduce the potential for information compromise on an enterprise scale.

7.2 Security Risk Management Uncertainty

Diversity in Nature can be a good thing or a bad thing. Genetic mutations can lead to adaptation but can also result in a species' demise. Diversity is the product of natural selection, where the latter creates opportunities for reproductive success. Diversity is part of Mother Nature's long-term risk management strategy.

Diversity in IT environments can also be beneficial or deleterious depending on the frame of reference. In general, attackers will be thwarted by IT environments with greater diversity since trial and error in identifying targets of opportunity is a frequent *modus operandi*. More diversity translates to a lower probability of successfully identifying a vulnerability or information asset of interest.

However, diversity in security risk management can represent a security problem. Ideally every risk factor is managed by at least one security control. A strictly probabilistic argument suggests if there are n risk factors, and the probability a risk factor is managed is x, the probability all risk factors are managed equals x^n assuming all risk factors are created equal. Therefore, the probability of risk management decreases exponentially as the number of risk factors increases. Note the probability no risk factors are managed equals $1 - x^n$ so that probability increases with an increasing number of risk factors.

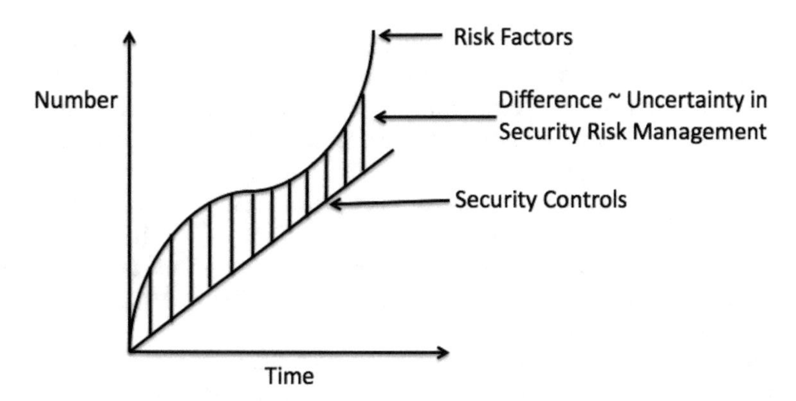

Fig. 7.1 Security risk management uncertainty first approximation

Of course, if the principal concern is whether a risk factor is managed by a security control or not it ignores whether a particular security control is actually adequate for the job. In the interpretation adopted herein success is strictly digital, where the probability of risk factor management equals one minus the probability of non-management.

As first mentioned in Chap. 5, the difference between the number of risk factors and security controls might be a simple if naïve indicator of cybersecurity risk. Specifically, it might indicate the magnitude of security risk management uncertainty. That magnitude would be expected to increase as the risk factors for information compromise increased relative to a fixed number of security controls.

Figure 7.1 depicts security risk management uncertainty as the difference between the number of risk factors and the number of security controls. This view of uncertainty clearly demands refinement but it will do for a first approximation.

Intuition suggests more expansive IT environments would result in greater security risk management uncertainty. Furthermore, a huge conceptual leap is not required to imagine such uncertainty contributes to the potential for information compromise. A stochastic approach to security risk management provides an analytic basis for such an assertion.

Consider an IT environment where N risk factors have been "measured" with respect to their security risk management status. If each risk factor has an uncertainty Δu_i that is random and independent, the total risk management uncertainty of the entire collection of risk factors (U) equals the square root of the sum of the individual uncertainties,[1]

$$U = \left[(\Delta u_1)^2 + (\Delta u_2)^2 + \ldots + (\Delta u_N)^2\right]^{1/2} \tag{7.1}$$

[1] John R. Taylor, *An Introduction to Error Analysis; The Study of Uncertainty in Physical Measurements*, University Science Books, 2nd Edition, Sausalito, CA, 1997.

The implication of Eq. (7.1) is the aggregate uncertainty in IT environments is scale-dependent if individual risk factor uncertainties are random and independent.

For example, assume an IT environment contains 100 elements, and each element contains a single risk factor for information compromise. If the uncertainty associated with each risk factor equals Δu, which is admittedly unlikely if uncertainties are indeed random, the uncertainty of the entire environment relative to an environment containing only one risk factor equals $[100(\Delta u)^2]^{1/2}/[(\Delta u)^2]^{1/2} = 10$.

If the IT environment contains 10,000 equivalent elements, the ratio now yields 100. Increasing the scale of the IT environment by 100 increases the ratio of uncertainties by a factor of 10.

Therefore, in the (very) special and unlikely case where security risk management uncertainty is the same for each risk factor across an IT environment, the total risk management uncertainty scales as the square root of the number of elements/risk factors that contribute to uncertainty.

Even if the individual uncertainties are not equal, the magnitude of U would still vary depending on how many segments of the IT environment are being assessed. However, in that case the simplification used in the above example would not be valid.

The point of specifying (7.1) is to show how assumptions of randomness in IT environments can have operational implications. If security risk management is assumed to be a stochastic process an estimate of the security risk management uncertainty *in aggregate* would be expected to scale with the size of the environment, i.e., the number of risk factors.

However, that fact by itself is not much use operationally. We have already acknowledged that accounting for each risk factor in assessing cybersecurity risk is infeasible. Therefore, (7.1) does little more than confirm that under the special condition where security risk management is a stochastic process/information source, the aggregate value of risk management uncertainty increases with more risk factors.

To be operationally useful we must somehow quantify this effect or at least provide a framework to evaluate IT environments and specify any assumptions and associated caveats. An indicative figure for complexity can be achieved by invoking the probabilistic approach to security risk management first described in Chap. 6. The objective of this chapter is to expand on that approach.

We now know information entropy quantifies the diversity of a message source, where a stochastic security risk management process qualifies as such a source. A specific probability distribution can be used to describe this process. Once such a model is adopted, the effect of message source diversity on IT environment unpredictability follows directly from information theory, and unpredictability is directly related to complexity as defined below.

7.3 Uncertainty, Diversity and Complexity

It is generally taken for granted that IT environments are complex. But the fact is many IT professionals might struggle to describe this condition let alone define it. Complexity in IT environments is reminiscent of the famous quote uttered by Supreme Court Justice Potter Stewart about pornography: "I can't define it but I know it when I see it."

Its ambiguity notwithstanding, complexity is a term often used to describe most commercial IT environments. However, a formal definition of this condition is essential if the objective is to identify its presence and to assess its effect on the potential for information compromise. One candidate for a definition might be the following simple statement,

> Complexity is the state or quality of being intricate or complicated.

To a first approximation this definition seems reasonable. But one problem is the terms 'complex' and 'complicated' are not equivalent although they are frequently conflated. 'Complicated' refers to a high degree of difficulty whereas 'complexity' is a function of the number of components or steps in an entity, system or process. Notably, the implication of complexity should be uncertainty rather than difficulty.

The path to a more precise definition of complexity might begin with describing its fundamental features. Namely, any complex entity, system or process must be multi-faceted and diverse. Both features must be present in order for an entity or process to qualify for complexity bragging rights.

For example, an entity consisting of 100,000,000,000 identical facets would not be considered complex. Equally, an entity with only three unique facets would not qualify for this designation despite its significant diversity. However, an entity comprised of 100 *unique* facets might indeed qualify as complex given its multi-faceted and diverse makeup. These examples suggest there is no defined threshold for complexity, but the goal is to identify a more objective if not quantitative criterion.

If both the number of facets and their diversity are required for an entity or process to qualify as complex, it seems reasonable to stipulate that the magnitude of complexity is inversely proportional to the probability of correctly guessing its correct configuration or sequence from its constituent elements or steps.

This statement is actually saying that complexity relates to the *unpredictability* associated with the possible configurations or steps, which is a function of the uncertainty/diversity of the guessing process. From this vantage one begins to see the connection between the diversity of a security risk management process/message source and the unpredictability of the results.

Describing a condition of maximum complexity will help describe complexity more generally. We again invoke an analogy with physical systems, which entails re-surfacing the concept of thermodynamic entropy, whose relationship with information entropy was discussed in Chap. 6.

The accessible states of a physical system were first introduced in Chap. 1. These states are the set of microstates that are consistent with the constraints imposed on the system, and are represented by the possible configurations of its elements, which in the case of a physical system are atoms or molecules.

If a constraint such as the total system energy changes, the number of accessible states will change, and the probabilities associated with the system being in a particular state will also vary. For example, if constraints are removed or relaxed, all the microstates formerly accessible to the system are still accessible in addition to a new set of states. Therefore, the probability the system will be found in one of the accessible states increases.

Assume there are two independent physical systems, where the number of accessible states of the first system is Ω_1 and the number of accessible states of the second system is Ω_2. Since the systems are independent, the total number of accessible states is given by $\Omega = \Omega_1 \times \Omega_2$. Since these are physical systems, the thermodynamic entropy is given by $S = k \ln(\Omega)$, where k is Boltzmann's constant. Therefore, S expresses the total number of accessible microstates of a physical system.

A maximally complex entity or process is loosely analogous to an isolated physical system in equilibrium. The latter is defined as the condition where all possible microstates are equally likely. By analogy, a condition of maximum complexity in a non-physical system might be one where all possible configurations of the individual elements have an equal probability of being correct. A key assumption is there is no a priori knowledge of the system that would bias an outcome.

Consider a drinking glass that breaks into two pieces after being dropped on the floor by a clumsy drinker. The process of reassembly is trivial. It hardly requires confirmation via a calculation but we will do it anyway for completeness. Here reassembly is assumed to be a random variable, and the probability of correctly guessing the reassembled state is 1. Therefore, the entropy $H = -\log(1) = 0$ bits/reassembly step since there is only one possible configuration of the two pieces. In other words it requires zero bits of information to describe the reassembly outcomes.

Now assume the glass has been smashed into a thousand unique pieces. The reassembly process is quantifiably more complex because the probability of successful reassembly via guessing is inversely related to the number of constituent pieces. Note the entropy decreases each time a piece is successfully glued in place since there are incrementally fewer pieces from which to choose. From these examples and the discussion in Chap. 6 it is now possible to specify a more precise definition of complexity as follows,

> Complexity is a condition characterizing multi-faceted entities, systems or processes where the probability of guessing the correct configuration or sequence of facets in the absence of a priori information about that entity, system or process is less than one. The lower the probability the more complex the entity, system or process.

Based on this definition, any entity or process consisting of multiple, diverse elements is complex, and the magnitude of complexity is determined by the uncertainty/diversity of the (random) selection process. The next step is to apply this definition of complexity to IT environments, pursuant to assessing cybersecurity risk on an enterprise scale.

7.4 A Cybercomplexity Scaling Relation

Recall the original motivation for information theory was to quantify the limits of electronic signal transmission. It turns out the amount of Shannon information that can be transmitted via a channel, i.e., the channel capacity (C), as measured in bits-per-second, is limited by noise.

The less bits required to be transmitted for a given message source the more efficient the signal transmission process. The information entropy (H) of an ergodic message source establishes the maximum rate of information transmission. For the interested reader, the maximum rate is given by C/H.[2]

The information theory formalism can be applied to IT environments. Specifically, if security risk management is assumed to be a stochastic process, we can view it as a message source. As described in Chap. 6, the diversity of this message source will determine the unpredictability of the resulting states. In this reimagined IT environment, security risk management is further assumed to be a binary stochastic process, where the two process outcomes are managed and unmanaged risk factors.

We know from Eq. (6.5) in Chap. 6 that the number of probable states resulting from a binary stochastic process is 2^{MH}. We also know from the discussion in Sect. 6.5 of that chapter that 2^{MH} information bits are required to characterize the spectrum of states, up to a maximum of 2^{M} bits. Recall H equals the process entropy specified in bits per process event and M is the number of events. All other states resulting from this process have negligible probability.

It is clear why this formulation is relevant to optimizing information transmission. It specifies the number of bits-per-symbol necessary to transmit the messages that are likely to occur based on the statistical characteristics of the message source, which can be used to optimize signal bandwidth.

In applying this formulation to IT environments, security risk management becomes a stochastic process with entropy (H) and risk management events correspond to the total number of IT environment risk factors (M). Therefore, Eq. (6.5) of Chap. 6 reveals it is risk management entropy *in combination* with the number of risk factors that drives the unpredictability of the resulting risk-managed states. Note as the magnitude of M times H increases, the unpredictability grows non-linearly because the number of probable states scales exponentially with M times H.

Since unpredictability equates to complexity according to the definition cited previously, the unpredictability as calculated above yields a metric for IT environment complexity. The latter condition is hereafter referred to as *cybercomplexity.*

The exponential dependence of complexity on M times H means small changes in their product can yield big changes in cybercomplexity. The operational implication

[2] Shannon called the C/H limit the fundamental theorem of the noiseless channel, which he described as follows: "Let a source have entropy H (bits per symbol) and a channel have a capacity [to transmit] C bits per second. Then it is possible to encode the output of the source in such a way as to transmit at the average rate of C/H - ε symbols per second over the channel, where ε is arbitrarily small. It is not possible to transmit at an average rate greater than C/H."

is that cybercomplexity grows *exponentially* as the scale of the assessed environment increases since more risk factors are likely to be present and risk management uncertainty would also be expected to increase.

This exponential dependence also suggests cybercomplexity is the principal contributor to cybersecurity risk on an enterprise scale. In fact, the competition for this title appears to be slim. Although proving its number one status might not be possible, contemplating the converse scenario might constitute a proof in reverse. Such a proof essentially follows the logic of the thought experiment discussed in earlier in the chapter.

If an IT environment only consists of two individuals who can only communicate with each other the opportunities for information compromise are nil. Security risk management uncertainty would be near zero and the number of risk factors is close to the same value. Such a small environment would definitely not require a CISO.

Now iteratively add *identical* elements to this IT environment. Nothing has substantively changed from a security risk management perspective. The IT environment is a self-contained ecosystem with no internal or external sources of security risk that would affect the potential for information compromise.

Even if the number of risk factors increased with additional elements, the increase would be linear since the number of risk factors per added element is constant. Most importantly, security risk management uncertainty is still zero since the probability distribution describing a stochastic security risk management process is unity with attendant implications to cybercomplexity.

This scenario is equivalent to the broken glass scenario discussed earlier in the chapter. In this example the individual pieces are identical. From a strictly complexity perspective it doesn't matter how many pieces result from the glass' demise if all constituent pieces are identical.

However, any deviation from this prescribed scenario has significant implications to complexity. Specifically, if non-identical elements are added, security risk management uncertainty is no longer zero. The effect of additional risk factors in the face of non-zero security risk management uncertainty exponentially increases with more elements. Any deviation from a non-zero security risk management uncertainty condition has a profound effect on the complexity of the ecosystem, especially if the number of risk factors is significant.

The key point is the specific types of elements being iteratively added to the environment is irrelevant unless the devices added are not the same type of device; the diversity associated with a multi-faceted environment coupled with the number of devices drives complexity. Therefore, the significant change in the risk profile results from complexity since the effect is scale-dependent.

Therefore, the status of complexity as the principal contributor to cybersecurity risk on an enterprise scale seems plausible after considering the previous thought experiment coupled with the non-linear dependence on M times H. The challenge is to imagine any other risk-relevant effect that would significantly increase the potential for information compromise *at that scale.*

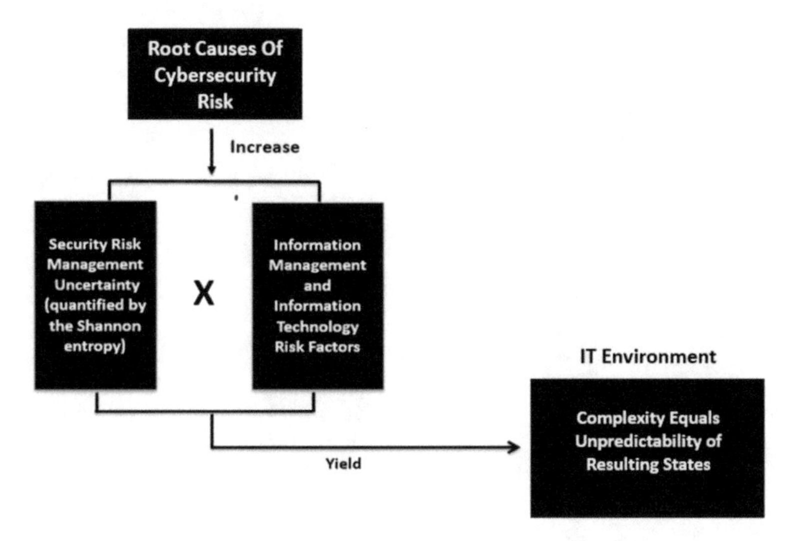

Fig. 7.2 The genesis of cybercomplexity

This conclusion also comports with common sense although common sense alone would not predict the non-linear scaling with IT environment size. Establishing that relationship required the adoption of a stochastic security risk management model.

Perhaps most importantly, the exponential scaling relation describing cybercomplexity points to the types of security controls that are required to reduce IT environment complexity. These controls are effective because they impose uniformity, which is an antidote IT environment unpredictability, i.e., cybercomplexity.

Finally, a brief word is required concerning the benefits and limitations of a model based on classical information theory.

The power of the model is in enabling a macroscopic view of IT environments by ignoring the status of individual risk factors. However, this model is not predictive. It cannot specify a required security control setting or how many security controls are necessary to reduce cybercomplexity to a particular value. It describes the statistical features of IT environments that result from a very specific and idealized security risk management process.

The objective is to leverage this process and thereby gain insight into the drivers of cybercomplexity while cautiously generalizing to more deterministic threat scenarios. One must recognize the limits of such generalizations but equally not ignore the potential lessons derived from even highly scripted scenarios.

Figure 7.2 illustrates the genesis of cybercomplexity.

Chapter 8
Cybercomplexity Metrics

8.1 Introduction

It is worth reiterating that commercial IT environments typically contain myriads of elements, and each element likely contains risk factors for information compromise. The presence of numerous risk factors in even modest IT environments suggests any realistic threat scenario is characterized by significant complexity.

Describing the contribution of each element to the magnitude of cybersecurity risk is non-trivial if a brute force approach to assessing cybersecurity risk is adopted. In attempting to assess the aggregate effect of all risk factors, evaluating the status of each element would be a daunting task. Assessing cybersecurity risk on an enterprise scale is intractable if not downright infeasible if such an approach is adopted especially since the risk factors can fluctuate in time.

As revealed in Chap. 7, a potential way forward is to adopt a statistical approach to assessing cybersecurity risk. However, two assumptions are required to pursue such an approach. The first assumption is that IT environments are comprised of generic risk factors and security controls. The second is to assume security risk management is a stochastic process. These assumptions lead to generic states consisting of unique combinations of managed and unmanaged risk factors. The unpredictability of the resulting states determines IT environment complexity.

Various metrics emerge from this approach and these are discussed in this chapter. Despite their quantitative patina, the metrics do *not* facilitate the calibration of security controls thereby enabling proportionate adjustments to cybercomplexity. However, they can potentially facilitate comparisons of complexity across IT environments. Most importantly, the model and associated metrics point to the efficacy of specific types of security controls as described in Chap. 10.

However, caveats apply to all such metrics. The model is based on a narrow and idealized form of security risk management, and therefore any metrics so derived must be interpreted with nuance and applied with appropriate caution.

C. S. Young, *Cybercomplexity*, Advanced Sciences and Technologies for Security Applications, https://doi.org/10.1007/978-3-031-06994-9_8

8.2 Absolute Complexity

By definition, complexity relates to the challenge in guessing a particular configuration of elements that comprises some entity or process. The complexity of an IT environment also relates to the magnitude of this challenge. Specifically, the magnitude of cybercomplexity is inversely proportional to the probability of guessing a particular configuration of managed and unmanaged risk factors, i.e., a specific state resulting from security risk management.

Specifically, the *absolute complexity* is given by the following scaling relation, where this form of C should not be confused with the correlation-time function discussed in Chap. 3,

$$C = 2^{MH} \tag{8.1}$$

Expression (8.1) is identical to expression (8.5) in Chap. 6, i.e., the number of probable states resulting from any binary stochastic process/message source.

The inverse of C yields the *unpredictability* or magnitude of uncertainty in guessing a particular IT environment configuration. In other words, C^{-1} specifies the probability an IT environment exists in a specific probable state following security risk management. The reader is once again reminded each state consists of a unique combination of managed and unmanaged risk factors.

Since the information or message source in this case is a binary stochastic risk management process, at the extreme values of H, i.e., zero and one, IT environment unpredictability is a minimum and maximum, respectively. It is again worth noting that C scales exponentially with the product of M and H. The scaling relation implies C^{-1} decreases at the identical rate.

If probable states result from a stochastic risk management process there must also be improbable states. The latter are states where the probability resulting from the stochastic risk management process is negligible. If 2^{MH} equals the number of probable states, the number of improbable states equals $2^{M} - 2^{MH}$. Importantly, C^{-1} is only a function of the number of probable states.

Figure 8.1 illustrates the details of the absolute complexity metric and the corresponding metric for IT environment unpredictability, C^{-1}.

M can assume any value since the number of risk factors is potentially unlimited. In contrast, the entropy of a binary stochastic process is constrained to vary between zero and one. The implication is that cybercomplexity will be large even when H is low if numerous risk factors are present. Of course, if H is zero for a given IT environment, which implies the stochastic process message source has zero diversity, cybercomplexity is always a minimum.

Recognize the decision to model the risk management process as a binary probability distribution is arbitrary. One could equally assume there are six risk management outcomes. In that case the risk management process would be analogous to the throw of a die rather than the toss of a coin. Unfortunately, if there are more than two outcomes the mathematics become messy, and any incremental insights

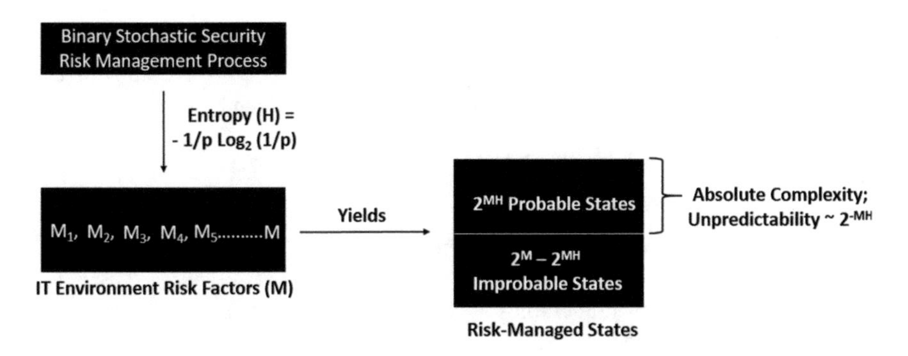

Fig. 8.1 Absolute complexity and IT environment unpredictability (cybercomplexity)

derived from a more involved model would be relatively insignificant. Non-binary risk management processes are discussed in Sect. 8.5 of this chapter.

Perhaps it is counterintuitive that cybercomplexity is expressed in bits. Recall information entropy equals the number of bits needed to specify each outcome of an information source. In the simple case of a binary stochastic process, where each outcome has equal probability, i.e., $p = 1 - p = 0.5$, a single bit is needed to specify a process outcome. That result was calculated in Chap. 6. When the probabilities are not equal, less bits per outcome are required, which translates to decreased entropy.

Perhaps a more intuitive way to think about information entropy is in terms of an expression of ignorance. Recall we alluded to this notion with respect to probability distributions in Chap. 6. The probability of identifying a particular element at random is lower for entities containing a larger number of elements. Therefore, we know less about more complex entities, and the information entropy as specified by the negative logarithm of the aforementioned probability compactly expresses the lack of knowledge, i.e., ignorance.

For example, if an entity consists of only two elements, there can be only one configuration of those two elements. Since $\log_2(2) = 1$, the logarithm expressed the fact there is zero diversity in a process or entity consisting of only two elements.

We can also think of information entropy in terms of the number of bits required to encode messages derived from message sources, which is how Shannon intended. For example, the probability of a specific outcome when there are only two possible outcomes of equal probability is obviously one half.

If the probabilities of a binary stochastic process/information source are equal, we know the information entropy is given by $-\log_2(1/p)$. Therefore, it follows that $-\log_2(1/2) = 1$.[1]

In other words, and in the parlance of information theory, it requires one information bit to encode a message generated by a binary message source when the probabilities of the two possible outcomes are equal.

[1] $-\log_p(1/p) = \log_2(1/1/2) = \log_2(2) = 1$.

Suppose the probability of a managed risk factor is now 0.3 instead of 0.5. Therefore, the probability of an unmanaged risk factor must be 0.7. Again invoking Eq. (8.2) of Chap. 6, the entropy for this probability distribution is given by,

$$\begin{aligned} H &= -\left[0.3 \times \log_2(0.3) + 0.7 \times \log_2(0.7)\right] \\ &= -[-0.52 - 0.36] \\ &= 0.88 \text{ bits/assessment outcome} \end{aligned}$$

Therefore, the biased process now requires only 0.88 bits to encode messages generated by this message source.

Now suppose the stochastic process is repeated numerous times. For example, one might toss a single coin 100 times or toss 100 coins only once. Recall these two processes are equivalent since a coin toss is an ergodic process. In either case the amount of information required to characterize the information source will scale according to the scaling relation specified by (8.1): the information bits per-outcome times the total number of outcomes.

If security risk management is a binary stochastic process/message source, there are 20 risk factors/risk management outcomes, and H equals 0.88 bits per risk factor, C is calculated as follows,

$$\begin{aligned} C &= 2^{\text{M risk management outcomes} \times \text{H bits/outcome}} \\ &= 2^{(20 \times 0.88)} \\ &\sim 310 \text{ bits} \end{aligned}$$

Therefore, 310 information bits are required to encode the spectrum of probable states resulting from security risk management applied to this IT environment. Next consider the number of probable states when H equals one. The absolute complexity (C) now equals $2^{MH} = 2^{(20 \times 1)} = 1{,}048{,}576$, which corresponds to the number of bits required to encode the maximum number of probable states in this instance.

It requires more information bits to characterize an information source when risk management uncertainty is a maximum assuming the number of risk factors remains constant. When H equals one, the IT environment is characterized by maximum unpredictability and hence maximum complexity since C^{-1} is a minimum.

Note the same value of complexity results if any two complementary probabilities are reversed. For example, suppose $p = 0.7$ for a managed risk factor and $1 - p = 0.3$ for an unmanaged risk factor. Although the resulting probable states would now contain a predominance of managed risk factors, which presumably reduces the potential for information compromise, the absolute complexity would be the same as it is when p(managed) = 0.3 and p(unmanaged) = 0.7. This result is a by-product of the symmetry of a binary stochastic process, which is immediately evident from Fig. 6.9.

It is worth repeating that the power law scaling relation means the number of probable states can be astronomically large in IT environments containing even a modest number of risk factors. This condition is especially true when the entropy is high. Such a condition might seem counterintuitive because of the connotation of the word "probable." The point is each probable state counts in the calculations of C and C^{-1}, and each probable state is equally probable. Therefore, the magnitude of unpredictability and hence cybercomplexity scales accordingly.

Figure 8.2 graphically illustrates the non-linear dependence of C on the product of M and H for two IT environments, where M = 5 and M = 10. The difference in the rate of growth is striking, even for IT environments containing a relatively small number of risk factors.

Figures 8.3 graphically illustrates the rapid growth of C with only small increments in risk management entropy for an IT environment containing 1000 risk factors. The y-axis is logarithmic (base 10), and therefore we see C increases astronomically for relatively modest increases in entropy.

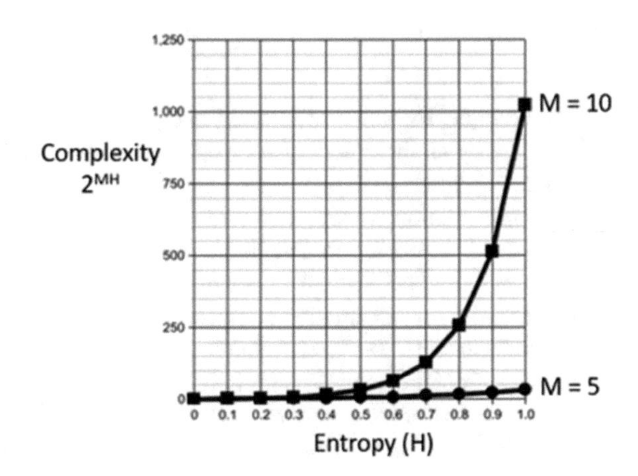

Fig. 8.2 Non-linear dependence of cybercomplexity on entropy and the number of risk factors

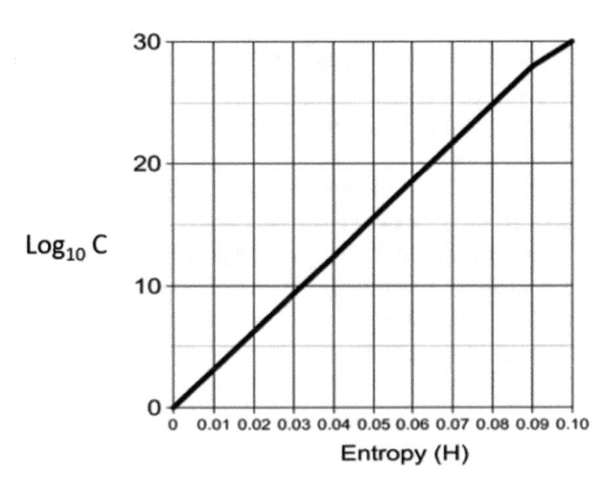

Fig. 8.3 Growth in C with small increments of H (1000 risk factors)

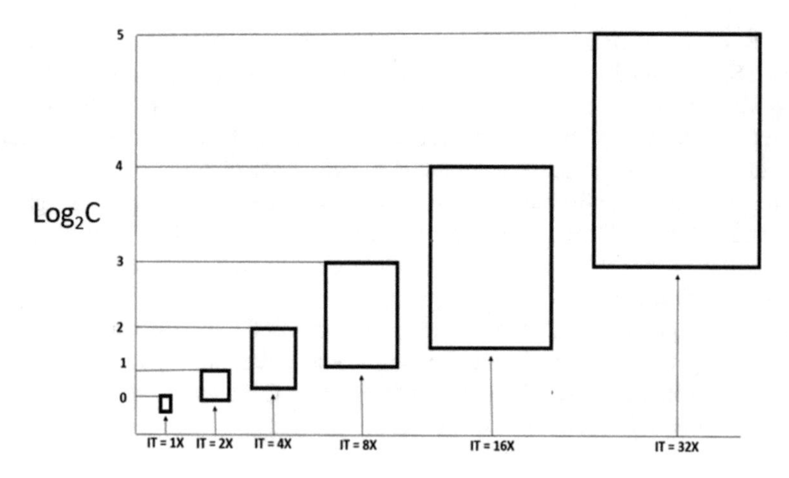

Fig. 8.4 Graphical representation of non-linear growth in absolute complexity

For example, the value of C when H equals 0.04 is 2^{40} or 1.1 trillion bits. When H increases by an increment of only 0.01 to a value of 0.05 C becomes 2^{50} or 3.2×10^{15}, i.e., 3.2 quadrillion bits.

Note information entropy is a relative figure since it is impossible to precisely calibrate H. Therefore, and despite its name, C and C^{-1} are relative quantities. For example, it would be difficult to meaningfully interpret 3.2 quadrillion bits in terms of the potential for information compromise. We will discuss issues associated with the calibration of H later in this chapter.

Figure 8.4 illustrates the growth in C each time the IT environment size is doubled, where the IT environment is represented by $2\times, 4\times, 8\times$ etc., noting H is held constant. This graphic is meant to visually represent how cybercomplexity scales with the size of the IT environment.

The key take-away from the previous discussion is that the magnitude of cybercomplexity is scale-dependent. Therefore, focusing on individual vulnerabilities will likely not reflect the contributions to cybersecurity risk that are only manifest on an enterprise scale. Of course the converse is also true. The upshot is the following pronouncement:

The scale of a cybersecurity risk assessment must be matched to the scale of the vulnerabilities or IT environment requiring assessment.

The absolute complexity offers potentially useful insights into cybersecurity risk. Such insights specifically relate to comparisons of complexity across IT environments, i.e., relative complexity. Metrics relating to relative complexity are the subjects of the next two sections.

8.3 Relative Complexity

The absolute complexity metric (C) specifies the number of probable states resulting from a stochastic risk management process applied to an IT environment. Notably, this metric ignores the states that are *unlikely* to result from a stochastic security risk management process, i.e., the *improbable* states.

As its name implies, each improbable state has a vanishingly small probability of occurrence. Nevertheless their number also grows exponentially with an increasing number of IT environment risk factors. This fact is important in order to appreciate the relative complexity metric as described immediately below.

The relative complexity metric (C_r) of an IT environment is defined as the ratio of the maximum number of probable states to the number of measured or actual probable states, i.e., $2^M/2^{MH}$. This ratio represents the *deviation* from a maximum entropy condition, which is easily seen by recalling $2^M/2^{MH} = 2^{(M-MH)}$.

If M is large the absolute complexity will be astronomically large due to the exponential growth caused by the product of M and H. But such a condition also implies the deviation from a maximum entropy condition will be similarly astronomical, even for low values of H. By comparison, an IT environment containing many fewer risk factors but the same value of H will be closer to a maximum complexity condition despite having a much lower value of absolute complexity.

Consider two IT environments with 10 and 1000 risk factors, where H is estimated to be 0.2 bits per risk factor in both environments. In the IT environment with 10 risk factors the absolute complexity, i.e., the number of probable states, can be encoded with $2^2 = 4$ bits. In stark contrast, the absolute complexity of the IT environment containing 1000 risk factors requires $2^{200} = 1.6 \times 10^{60}$ bits for encoding.

This figure is astonishingly large.[2] However, the relative complexity of the environment with 1000 risk factors equals 2^{800}, an even more astonishing number. In contrast, the IT environment with 10 risk factors has a relative complexity equal to $1024/4 = 256$. Therefore, the smaller environment is closer to a maximum complexity condition despite being significantly less complex than the IT environment containing 100 times as many risk factors.

To reiterate, a larger value of relative complexity means the IT environment is further from the maximum complexity condition, 2^M. This phenomenon is explained as follows. IT environments with a large number of risk factors and a risk management entropy greater than zero but less than one also contain an exponentially large number of improbable states. The improbable states disproportionally subtract from a maximum complexity condition as the number of risk factors increases.

The number of improbable states in an IT environment with absolute complexity 2^{MH} is given by $2^M - 2^{MH}$ (d). Figure 8.5 shows the behavior of d for an IT environment with 10 risk factors as a function of H. Note the number of improbable states is zero when H equals one, i.e., a maximum. When the entropy is a maximum the

[2] It is estimated there are between 10^{78} and 10^{82} atoms in the known, observable universe (www.universetoday.com).

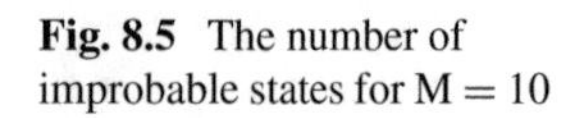

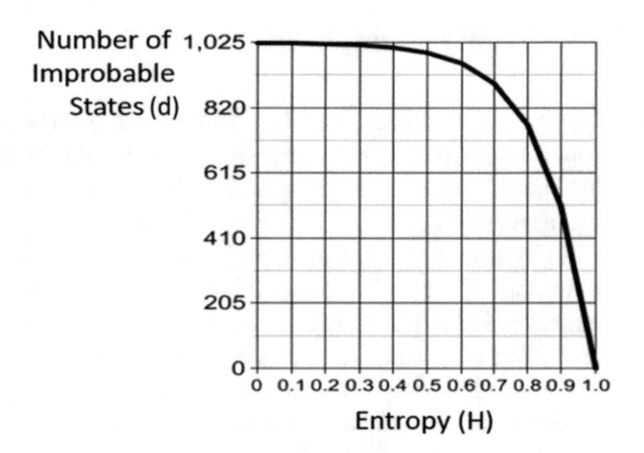

Fig. 8.5 The number of improbable states for M = 10

number of probable states must also be a maximum. Therefore, there are no improbable states since d measures the difference from a maximum complexity condition. When H equals one that difference is zero.

Figure 8.6 illustrates the relative complexity for two IT environments consisting of five and ten risk factors. Note using a linear scale to graph C_r is challenging even for small environments due to the exponential dependence on the product of M and H.

The logarithm (base 2) of C_r also reveals risk-relevant phenomena, and is trivially computed as follows,

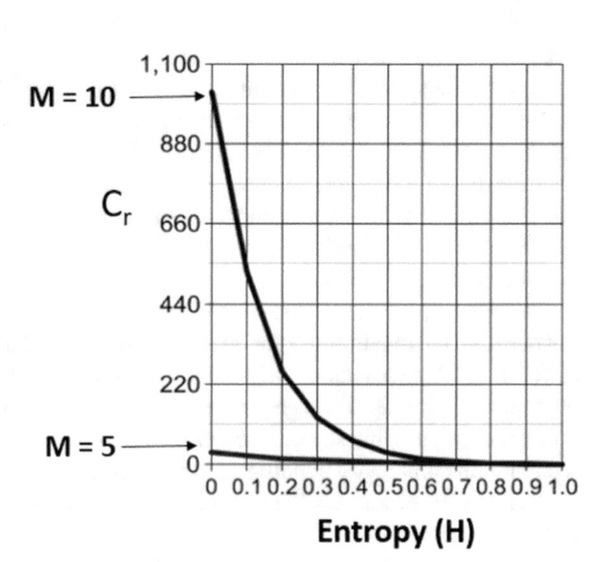

Fig. 8.6 Relative complexity of two IT environments

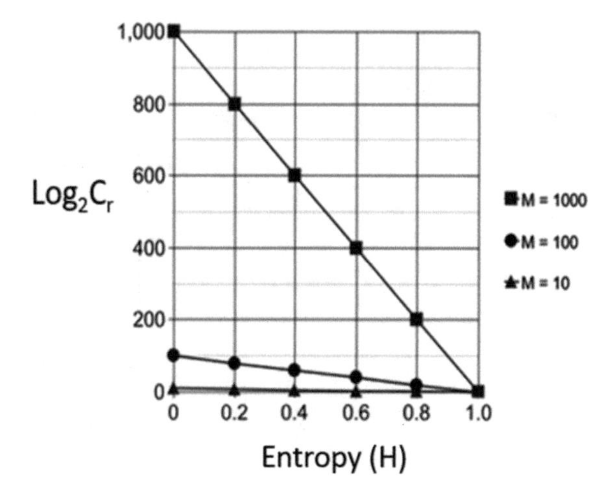

Fig. 8.7 Comparison of $\text{Log}_2\ C_r$ for three IT environments

$$\begin{aligned}
&\log_2 C_r \\
&= \log_2\left[2^M/2^{MH}\right] \\
&= \log_2 2^{M(1-H)} \\
&= M(1-H)
\end{aligned} \tag{8.2}$$

Note that $\log_2 C_r$ converges to zero with increasing values of H. Again, this condition arises because C_r expresses the deviation from a maximum complexity condition. Therefore, this metric must converge to zero as H approaches 1, the maximum value. C_r effectively normalizes an IT environment relative to its maximum complexity condition. This normalization facilitates comparisons of IT environments that contain varying numbers of risk factors.

Figure 8.7 plots $\log_2 C_r$ for IT environments containing 10, 100 and 1000 risk factors, as a function of risk management entropy.

We see the deviation from a maximum complexity condition is highest for the IT environment with 1000 risk factors and lowest for the environment with 10 risk factors. However, the magnitude of absolute complexity is much greater for the former environment.

It is also clear from Fig. 8.7 that the *rate* of decline of $\log_2 C_r$ is much steeper for IT environments with numerous risk factors. The ratio of the number of probable states to the maximum number of probable states becomes increasingly significant for IT environments with higher security risk management entropy, and the steepness is determined by the number of risk factors, M. Clearly scale matters when assessing complexity in IT environments.

Figure 8.8 plots $\log_2 C_r$ as a function of the number of risk factors and constant entropy H.

Perhaps surprisingly, Fig. 8.8 reveals the rate of increase and the values of $\log C_r$ are greatest for environments with lower entropy. Once again $\log_2 C_r$ is expressing the *deviation* from a maximum complexity condition. Therefore, it makes sense that

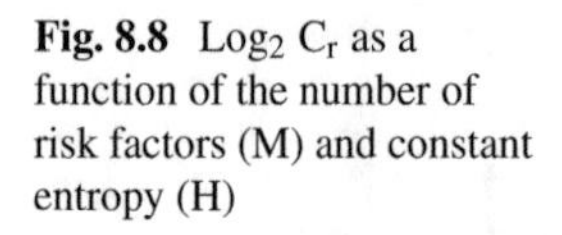
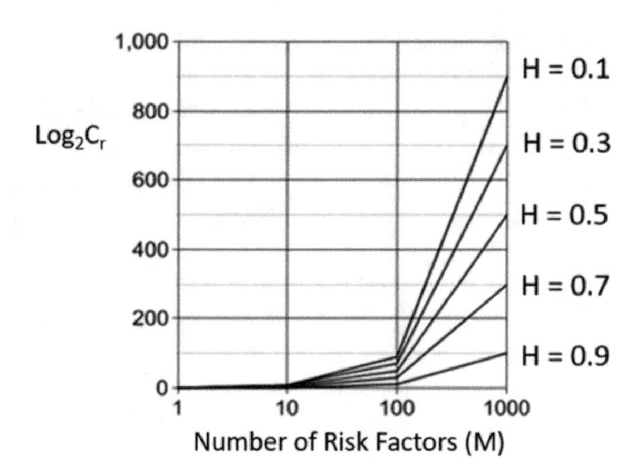

Fig. 8.8 $\text{Log}_2\ C_r$ as a function of the number of risk factors (M) and constant entropy (H)

both the relative complexity and its rate of increase are largest when the entropy is furthest from the maximum value.

We can write a more general metric for relative complexity. Instead of the difference between the "measured" complexity and a maximum complexity condition, one might specify the complexity relative to a specific value of entropy. The value of entropy might reflect some change in the IT environment, which would affect security risk management uncertainty.

For example, the metric might specify the security risk management uncertainty before and after a security control has been applied, which presumably affects the value of relative complexity. The form of the more general metric ($C_{r'}$) is as follows,

$$C_{r'} = 2^{MH}/2^{MH'} = 2^{M(H-H')} \tag{8.3}$$

In expression (8.3) H is the initial value of entropy and H′ is its value after a change to the IT environment has been implemented. We could once again take the logarithm of $C_{r'}$ without any loss of meaning, which means (8.3) reduces to M (H–H′).

In exact analogy with absolute complexity, $C_{r'}$ and log $C_{r'}$ reflect the deviation from a baseline or reference complexity condition. In this case the baseline is some parameter other than a maximum complexity condition. Using this metric to assess a particular risk-relevant process reveals its inherent limitations and alerts us to the dangers of misapplying cybercomplexity metrics.

Let's assume it is desired to know the effect of a security training program on the magnitude of cybercomplexity. We must first specify the value of information entropy that applies to the particular scenario of interest. In this case we choose the threat scenario to be phishing, which is a common form of social engineering. The standard *modus operandi* employed by an attacker is to induce a network user to click on a link embedded in an email, which connects the user to a malicious website.

The stochastic process/information source that describes the vulnerability to phishing has two possible outcomes: an embedded hyperlink link is clicked and the link is not clicked. Therefore, a binary stochastic risk management process seems applicable to this threat scenario and relative complexity can be easily calculated.

It has been anecdotally reported that following anti-social engineering training approximately one third of graduates still click on malicious links in emails. If the actual fraction is known, a calculation of $C_{r'}$ will yield the deviation from the baseline or reference complexity condition, which is assumed to be 0.33.

In other words, if 33% is the most optimistic click rate, that figure could be used as a baseline probability for hyperlink clicking. So if p(click) equals 0.33, p(not click) must equal 0.67. The entropy (H) is therefore 2.17 bits/click.

However, if the actual click rate following training is p(click) = 0.27, an apparently fantastic risk management result, it must follow that p(not click) = 0.73. Therefore, the entropy (H′) of the process following training equals 2.34 bits/click.

The potential for information compromise following training has clearly decreased since p(click) is lower than the expected value following training. However, the message source entropy has actually *increased*. Therefore, the potential for information compromise has improved but the magnitude of complexity resulting from this process has worsened.

The lesson here is entropy is agnostic to the polarity of the two probabilities. In other words, from an information theoretic perspective the respective probabilities have no inherent value. Analogously, the probability of a heads outcome is neither good nor bad relative to a tails outcome in a coin toss. Cybercomplexity is strictly a numbers game, where the entropy simply measures the diversity of the message source, and message source diversity has nothing to do with the relative goodness or badness of the two risk management outcomes.

From a complexity perspective the situation following training has indeed gotten worse. The following calculation reveals the relative increase in deterioration.

If 1000 emails is the volume of email traffic in a unit interval of time, and each email is assumed to contain a malicious hyperlink, the logarithm of the relative complexity increases by a value of 170 bits due to training. Note the absolute value of $H - H'$ ensures a positive value of the exponent,

$$\begin{aligned}
&\log_2 C_{r'} \\
&= \log_2 2^{M|H-H'|} \\
&= \log_2 2^{1000(|2.17-2.34|)} \\
&= 1000 \times (.17) \\
&= 170\,\text{bits}
\end{aligned}$$

C_r might actually be a more informative metric in this instance. Recall the reference entropy for C_r is a maximum complexity condition, i.e., $H = 1$. When H equals one the probability of clicking is the same as the probability of not clicking. Therefore, the process is equivalent to a fair coin toss. The predicate for using C_r in this context is the assumption that *any* deviation from a maximum complexity condition represents an operational improvement.

In other words, the assumption, which is subject to debate, is that a scenario resulting from a completely random process is operationally worse than one where the number of probable states consisting of a majority of "click" outcomes outnumbers states consisting of a majority of "no-click" outcomes.

Next consider the scenario where p(click) = 0.67 and p(not click) = 0.33. In this instance the two probabilities are reversed from the scenario described previously. However, the entropy equals 2.17 bits/click in both cases! This potentially confusing scenario is yet another result of the symmetry of binary stochastic processes. It again highlights the fact that complexity and the potential for information compromise are not necessarily related.

The universal implementation of two-character passwords is a prime example of the potential disconnect between the magnitude of information entropy in IT environments and the potential for information compromise. In summary, one must be clear about the precise meaning of information entropy as well as realistic about the operational relevance of complexity metrics in a particular context. The bottom line is all security controls should be simultaneously evaluated with respect to their effect on complexity and other possible avenues of information compromise.

8.4 The Density of States

Another risk-relevant metric is the density of states, which can assume several forms. The first version equals the ratio of the number of improbable states to the maximum number of probable states. This metric is similar to relative complexity except here the focus is on the *ratio* (and hence a density) of the number of improbable states to the maximum number of probable states 2^M, i.e., $(2^M - 2^{MH})/2^M$.

The metric merely confirms the fact that IT environments can contain many improbable states but still not exhibit high absolute complexity. For values of entropy near one, there is little to no difference between the number of improbable states and the maximum number of probable states, i.e., 2^M.

The density of states (D) can be rewritten as follows,

$$\begin{aligned} D &= \left(2^M - 2^{MH}\right)/2^M \\ &= 1 - 2^{MH}/2^M \\ &= 1 - 2^{M(H-1)} \end{aligned} \tag{8.4}$$

Note the density of states converges to zero when the information entropy approaches one. This condition makes sense since D also measures the deviation from a maximum entropy condition. If a maximum information entropy condition characterizes security risk management in an IT environment, the deviation from a maximum condition must equal zero.

The falloff in D for low values of H is slow, but at some critical value the falloff becomes precipitous. The effect is magnified for scenarios with a large number of risk factors. This phenomenon arises because the product of M and H is disproportionately amplified for large values of M, but H is constrained to vary between zero and one.

Furthermore, the ratio of the number of improbable states ($2^M - 2^{MH}$) to the maximum number of probable states (2^M) decreases steeply as H approaches a maximum value. The steepness is affected by the IT environment scale, i.e., the total number of risk factors present in the environment.

Figures 8.9 and 8.10 illustrate the behavior of the density of states for IT environments with 10 and 100 risk factors, respectively, and as a function of H.

There is a related but slightly different version of the density of states, which is denoted by D′,

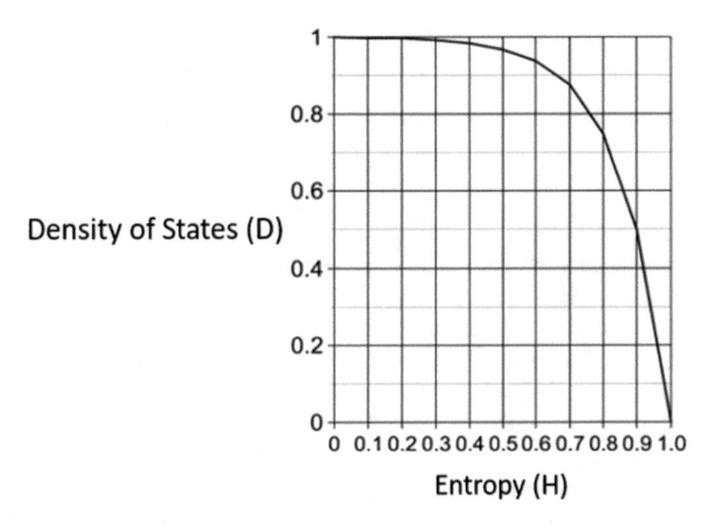

Fig. 8.9 Density of states for M = 10

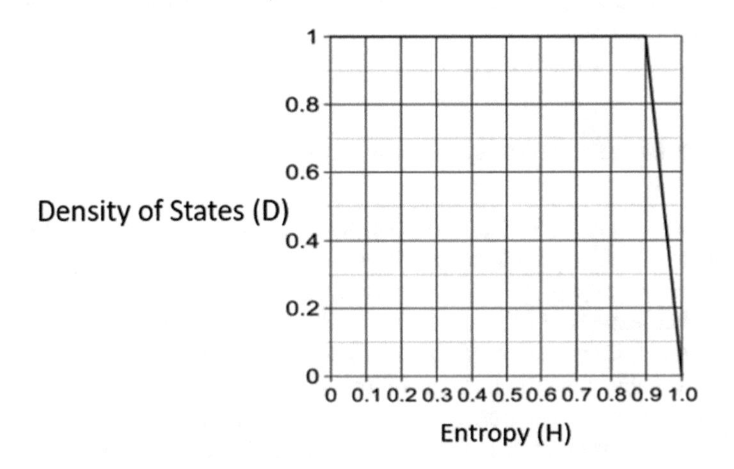

Fig. 8.10 Density of states for M = 100

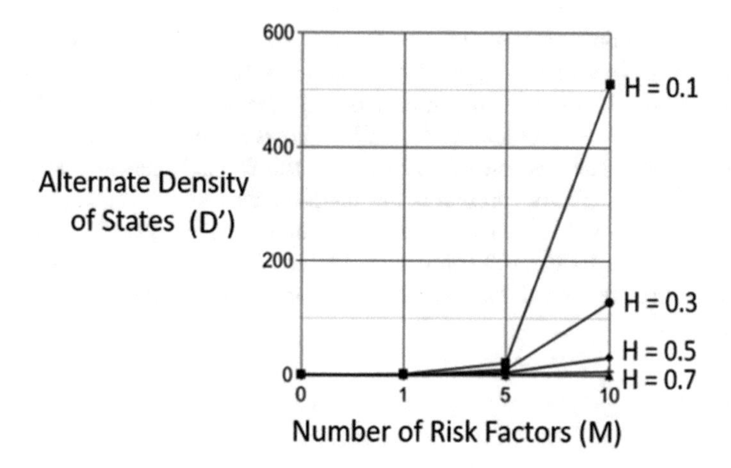

Fig. 8.11 Alternate density of states (ratio of the number of improbable to probable states)

$$D' = \left(2^{M} - 2^{MH}\right)/2^{MH} \tag{8.5}$$

This metric equals the ratio of the number of improbable states to the number of probable states. As is the case with D, the number of improbable states can be huge if M is large. But we know the number of probable states will also be large in such instances. The ratio of the two figures is particularly large for lower values of entropy. However, that ratio diminishes as the entropy decreases, and it actually converges to zero as H approaches one, i.e., a maximum entropy condition.

D′ once again expresses the deviation from a maximum complexity condition in IT environments. It reinforces the point that this deviation can be huge despite a high value of absolute complexity. Each curve in Fig. 8.11 represents the change in D′ for values of M between zero and ten and for constant values of H.

Table 8.1 summarizes the cybercomplexity metrics for IT environments containing M risk factors.

8.5 Non-binary Security Risk Management

The choice of a binary stochastic risk management process is completely arbitrary. It was selected exclusively because the entropy calculations were simple and therefore facilitated easy analysis. With an increasing number of outcomes, the number of possible combinations of probabilities that sum to one becomes overwhelming.

For example, if six risk management outcomes are possible instead of two, the number of combinations of six probabilities that sum to one would be difficult to express. Therefore, the spectrum of IT environment conditions would be correspondingly difficult to represent. In the case of six risk management outcomes the security risk management process is equivalent to the toss of a fair die, and once again Eq. (8.2)

Table 8.1 Summary of cybercomplexity metrics

Risk management outcomes (A, B) probabilities	Entropy (H)	Unpredictability	Absolute complexity (C)	Relative complexity (C_r)	Density of states (D)
Metric	$-\sum p \log p$	2^{-MH}	2^{MH}	$2^{M(1-H)}$	$1 - 2^{M(H-1)}$
p(A) or p(B) = 0	0	Minimum	Minimum complexity = 2° = single state	Maximum	Approximately 1 (maximum)
0 < p(A) and p(B) < 1; p(A) and p(B) ≠ 0.5	0 < H < 1	Intermediate	2^{MH} probable states	Decreases with increasing entropy	Decreases with increasing entropy
p(A) = p(B) = 0.5	1	Maximum	Maximum complexity = 2^{M} = maximum number of probable states	Minimum (converges to 1)	0 (minimum)

of Chap. 6 can be invoked to calculate the entropy,

$$
\begin{aligned}
H &= -\sum p_i \log_2(p_i) \\
i &= 1 \\
&= -\log_2(1/6) \\
&= 2.58 \text{ bits/outcome}
\end{aligned}
\tag{8.6}
$$

Figure 8.12 depicts the entropy of risk management processes with uniform probability distributions ranging from one to ten outcomes.

For a fixed number of risk factors and constant information entropy, the complexity significantly increases for processes with a greater number of outcomes. This result is actually immediately apparent from the form of the absolute complexity metric.

For example, the number of probable states resulting from security risk management processes consisting of two versus six outcomes equals 2^{MH} and 6^{MH}, respectively. The effect of multiple security risk management process outcomes on cybercomplexity is significant, and it results from the change in base, i.e., from two to six. Figure 8.13 illustrates the effect of the number of risk management outcomes on log C as a function of entropy, where M = 10.

Two points are worth noting. The first is to recognize the scale of the vertical axis in Fig. 8.13 is logarithmic (base 10). Therefore, the values of cybercomplexity for larger values of H are exponentially larger despite appearances.

Fig. 8.12 Information entropy as a function of risk management outcomes

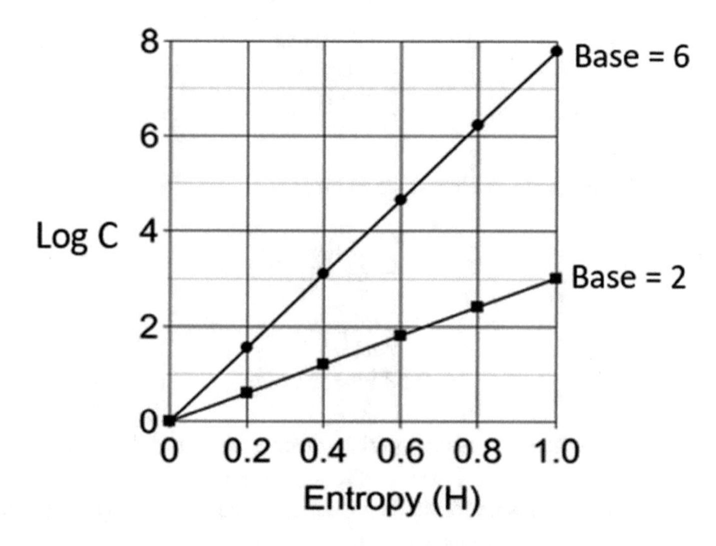

Fig. 8.13 Cybercomplexity for risk management processes with two and six outcomes (10 risk factors)

For example, if the entropy is 1, the linear value of cybercomplexity with six possible outcomes and 10 risk factors is 60,466,176 bits. In startling contrast, the cybercomplexity is only 1024 bits if there are only two outcomes for the same number of risk factors. The entropy is approximately 4.7 bits/outcome when six outcomes are possible whereas it is only 1 bit/outcome when just two outcomes are possible. The effect of exponential growth can be dramatic.

The key take-away is the effect of non-linearity on cybercomplexity is even more pronounced for non-binary risk management processes. Nevertheless, the fundamental results apply, irrespective of the number of process outcomes.

8.6 Information Entropy Calibration

As discussed in Chap. 1, there is no avoiding the fact that a reference is required to make *calibrated* adjustments to security controls. In the absence of calibration it is impossible to precisely reduce the magnitude of cybercomplexity in accordance with the tolerance for risk no matter how judiciously security controls are applied.

Calibration could theoretically be achieved in at least two ways. The first is through data collected from multiple IT environments. If it were possible to collect risk-relevant data, ideally from a statistically significant number of IT environments, it might be possible to correlate a particular IT environment change with changes in the value of security risk management entropy, and thereby estimate the magnitude of cybercomplexity.

The second method of calibration would be via experimentation. If a risk-relevant parameter could be isolated and then varied, the response to only these variations could be measured. The measurement would yield the magnitude of an entropy change relative to the change in the parameter.

Unfortunately both approaches are operationally challenging. It might be impossible to isolate a specific feature of an IT environment and thereby ensure the observed effect only relates to the parameter of interest. Furthermore, experiments in this context are unrealistic, in part because designing a protocol to include a statistical control is impractical.

The inability to calibrate security controls represents a general obstacle to precisely managing cybersecurity risk. However, this condition does not imply it is impossible to identify security controls that have an ameliorating effect on the magnitude of security risk. For example, there is no need for calibration to conclude that multi-factor authentication is effective. It is just not possible to quantify the magnitude of the effect.

The point of this discussion is not to diminish the utility of identifying a simple model of cybercomplexity. A stochastic formulation of security risk management enables *inferences* regarding the magnitude of cybercomplexity. Most, if not all modern IT environments have numerous risk factors and therefore almost certainly exhibit cybercomplexity. The operational benefit of a model of cybercomplexity is its potential to inform practitioners how to address this condition.

The presence of certain organizational features that tend to amplify the number of risk factors and/or increase risk management uncertainty suggests conditions are ripe for cybercomplexity. These features can be relatively easy to identify if there is a predisposition to look for them. Cybercomplexity metrics might help confirm the effect of remediation efforts.

However, the most compelling result derived from cybercomplexity metrics is confirmation of the requirement for security controls that act macroscopically within IT environments. Such controls increase uniformity on an enterprise scale and therefore have the most significant impact on the root causes of cybercomplexity. These root causes and macroscopic security controls are discussed in the final section of this text.

Part IV
Cybercomplexity Genesis and Management

Chapter 9
Cybercomplexity Root Causes

9.1 Introduction

The objective of cybersecurity risk management is to identify and address the risk factors for information compromise or information-related business disruption via *the application of security controls in accordance with an organization's tolerance for risk.*

Although addressing cybersecurity risk factors via the application of security controls represents the crux of security risk management, the reality is financial and/or staffing resources are frequently insufficient to effectively address all risk factors in an IT environment. In fact, merely identifying every risk factor is frequently problematic.

Therefore, the prioritization of security controls is required, which is the purpose of a security risk assessment. Determining the risk factors most deserving of attention will require assessing all three components of risk in consideration of the financial and operational burdens imposed by identified solutions.

For example, the vulnerability component of risk associated with a nuclear missile attack is extraordinarily high. However, the resources required to successfully address a nuclear threat are significant. Presumably the low likelihood of such an incident justifies not addressing this threat. Such trade-offs represent the essence of security risk management.

Presumably every security risk manager hopes to identify the root causes of security incidents in order to prevent future incidents. An analogy with cardiac care would be implanting a stent versus implementing lifestyle changes. The former might be necessary to address an acute blockage of a coronary artery. However, addressing the root cause of the blockage frequently requires lifestyle changes especially if an infarction, stroke and/or future stents are to be avoided in the long-term.

At a high level, the root causes of cybercomplexity relate to organizational features that promote inconsistency, delinquency, incompleteness and incoherence in security risk management. These features also signal the presence of cybercomplexity since they tend to enhance risk management uncertainty and/or spawn risk factors for

C. S. Young, *Cybercomplexity*, Advanced Sciences and Technologies for Security Applications, https://doi.org/10.1007/978-3-031-06994-9_9

information compromise. In general, the root causes of cybersecurity risk ultimately relate to the *approach* to security risk management and in particular the integration of cybersecurity and risk-relevant processes, workflows and technologies.

The risk-relevance of addressing the root causes of cybercomplexity is threefold. First, and as discussed in Chaps. 7 and 8, an accretion of risk factors in conjunction with risk management uncertainty has a non-linear effect on the magnitude of cybercomplexity. Second, if the root causes of cybercomplexity are ignored, the conditions that make the environment ripe for information compromises will persist. Third, many of the root causes highlighted herein are related. Therefore, addressing one root cause will potentially have a salutary effect on others, which could amplify the effects of remediation.

The traditional approach to assessing cybersecurity risk is to focus on individual vulnerabilities. For example, much time and effort is spent remediating the results of automated network scans pursuant to addressing published vulnerabilities. Certainly each of these vulnerabilities increases the potential for information compromise to varying degrees. Addressing these vulnerabilities is necessary but not sufficient to manage cybersecurity risk since they are unlikely to address risk-relevant effects that are manifest on an enterprise scale. We note in passing these automated solutions frequently lack sufficient context to accurately assess the potential for information compromise in a given IT environment, despite ranking such vulnerabilities according to severity.

Perhaps the organizational feature with the greatest effect on cybercomplexity is the tolerance for risk. Although this tolerance is not a root cause of cybercomplexity *per se*, it reflects an organization's overarching philosophy toward security risk management. Moreover, it affects all aspects of security risk management because it determines the threshold for applying security controls to risk factors.

The tolerance for risk is a direct reflection of the organizational culture. An organization's culture has the most profound influence on cybersecurity risk since it determines the calibration point on the security risk continuum as depicted in Fig. 9.1. The tolerance for risk warrants inclusion in the panoply of root causes by virtue of its impact on the approach to security risk management and is therefore discussed next.

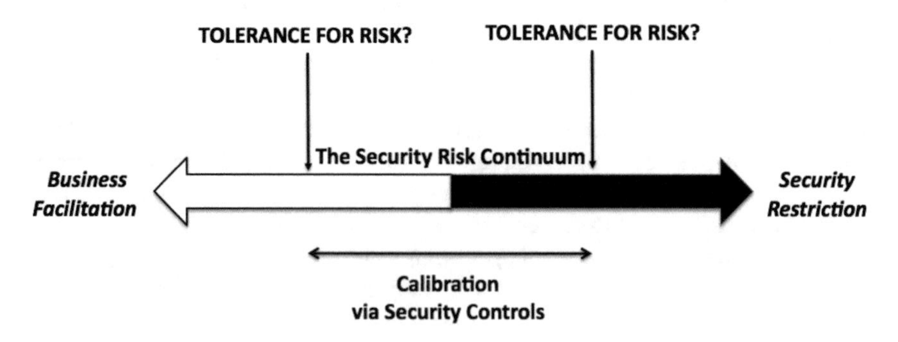

Fig. 9.1 Calibrating the security risk continuum

9.2 The Organizational Tolerance for Risk

As noted numerous times, the principal obstacle to precise security risk management is the inability to calibrate security controls and thereby adjust the magnitude of cybersecurity risk in accordance with some standard. That standard is the organizational tolerance for risk. The ongoing challenge is to determine and maintain that standard across the enterprise.

The organizational tolerance for risk exists on a theoretical continuum that is bracketed by two extremes. On one extreme is complete business facilitation, where any network communicator can do whatever is required to accomplish a particular business objective. At the other extreme is complete information restriction, where the principles of need-to-know, least privilege and information asset access denial by default are uncompromisingly rigid.

Figure 9.1 illustrates the security risk continuum and its calibration via security controls according to the organizational tolerance for risk.

Identifying the proper balance between security restriction and business facilitation across an enterprise can be non-trivial. The responsibility of determining where on the continuum an organization should exist rests with organizational leaders. The responsibility for ensuring the organization is appropriately balancing the tolerance as articulated by executive management is the IT security function. Note the balance will likely vary depending on the context, which adds to the enterprise-level security risk management challenges.

The tolerance for risk is an abstraction that can be difficult to articulate let alone quantify. Certainly an ambiguous tolerance for risk can promote incompleteness, inconsistency, exceptions, delinquency and/or incoherence in security risk management, particularly on an enterprise scale. Evidence of an ambiguous, irrational or mysterious tolerance for security risk should elevate concerns that cybercomplexity is enhanced.

Excessive permissiveness represents the hallmark of an organization's acceptance of an enhanced potential for information compromise. Examples of such permissiveness include the proliferation of administrative rights and/or weak log-in credentials, where both conditions would be expected to spawn numerous risk factors for information compromise. The approach to network segregation is another useful barometer for an organization's tolerance for risk.

A general condition of permissiveness-by-default is highly risk-relevant.

A possibly far-fetched indicator of such a condition would be a firewall setting of "permit ip any any," which allows all traffic from any source on any port to any destination. A firewall so configured is almost certainly reflective of an organization at odds with the principles of need-to-know and zero trust.

This firewall configuration would reflect an unusually high tolerance for risk according to any reasonable standard of security risk management. It is not difficult to imagine how operating at the extreme end of the continuum in Fig. 9.1 could expand the number of risk factors for information compromise and/or increase risk management uncertainty, and therefore have a significant effect on cybercomplexity.

Finally, one might reasonably ask how the organizational tolerance for risk is actually determined. After all, it is unrealistic to demand an organization adhere to a standard that is beyond its capability to assess. Unfortunately, there is no algorithm for determining the organizational tolerance for risk. Furthermore, the tolerance could change depending on the issue, location and even the timing of an assessment.

At a minimum, assessing the tolerance for risk requires examining all three components of risk relative to the resources required to manage each component. This juxtaposition is admittedly not prescriptive, and the results will be a function of the specific risk factor and other variables as noted above. Certainly the general permissiveness toward information access and authentication of identity are strong cybersecurity risk tolerance indicators.

The bottom line is determining and adhering to the tolerance for security risk is ultimately subjective, and it should be the focus of a centralized security governance process. However, such assessments are not devoid of objectivity. They require an evaluation of the business impact of security controls relative to the security risk management impact of business requirements for each component of risk associated with the potential for information compromise.

Specifically, cybersecurity risk assessments are required to determine if the application of security controls to risk-relevant processes, workflows and technologies facilitate business requirements without enhancing the potential for information compromise beyond some agreed threshold. Those thresholds must be established and enforced by arbiters with a global perspective of both security risk management and business requirements.

9.3 Convenience-Driven Culture

Culture is frequently cited as a risk-relevant organizational feature but its full impact on cybersecurity risk is not necessarily appreciated. Culture affects all aspects of organizational behavior and particularly the approach to security risk management. Therefore, the organizational culture can either be a highly effective security control or a significant contributor to the potential for information compromise.

It is worth explaining what is meant by culture in this context. The organizational culture is an orientation or mindset that is specific to each organization and/or organizational unit. This orientation is manifest as behaviors and processes that affect information management. Culture is an organizational bias that can result from explicit mandates or propagate via "osmosis" within and across generations of employees.

The effect of cultural forces supporting or opposing security risk management can be formidable. The direction and magnitude of these forces are frequently dictated by the security-convenience dialectic noted previously, which is actually a manifestation of the tolerance for risk. Every decision regarding the application of security controls to risk factors is ultimately a choice between implementing more or less security risk restriction versus facilitating more or less convenience.

Organizational culture can have a profound effect on cybercomplexity in particular. Risk factors and risk management uncertainty are natural by-products of a culture that allows, and perhaps even encourages, specific behaviors. The net result is enhanced unpredictability, which we now know characterizes cybercomplexity.

Cultural artifacts are frequently revealed to be the culprit in security incident *post mortems*. These incidents occurred because the culture has created conditions ripe for information loss or leakage. Inconsistent and/or liberal requirements regarding system authentication immediately come to mind. Again, such artifacts often result from the tension between security and convenience.

As noted in the previous section, the tolerance for risk mirrors the organizational culture. In fact, the organizational tolerance for risk and its culture are virtually synonymous. Adjusting the tolerance for risk is tantamount to changing the organizational culture, which will require active engagement by senior leadership.

In that vein, one noteworthy feature of all organizational cultures is they originate from the top and propagate downward, noting individual organizational units can possess mini-cultures especially if organizations are stove-piped. The culture frequently passes from one generation to another since leaders tend to pick future leaders that resemble themselves or at least agree with their way of thinking.

Finally, it should be emphasized that not all cultural artifacts are harmful. In fact, a significant challenge is to implement required cultural changes without destroying the cultural conditions that have historically contributed to the organization's success.

9.4 Structural and Functional Anomalies

The structure and function of an organization is frequently indicative of its culture, and at a minimum, can materially affect overall management, security risk management, communication, and information sharing. The organizational structure reflects the organizational priorities, and it tends to reinforce systemic issues related to security risk management.

Anomalies in the organization's structure and function can create conditions ripe for information compromise by increasing cybercomplexity. Risk factors for information compromise emerge from these anomalies, which tend to exacerbate incoherence and inconsistency in risk-relevant technologies, processes and workflows.

Examples of such anomalies are ambiguity in reporting lines, disconnects between business operations and security risk management, the exclusion of security best practices/principles in formulating and implementing business processes, workflows and technologies, and the isolation of key organizational entities. Security risk management uncertainty would be expected from an environment beset by such anomalies.

A high frequency of such anomalies and the resulting effect on coherent/consistent security risk management can ultimately be traced to the action or inaction of senior management and/or the board of directors. There is an explicit connection between the implementation of security controls on an enterprise scale and the individuals who establish the organizational structure, oversee business functions and establish the culture. Because the culture/tolerance for risk is established and sustained by the most senior executives, it is no exaggeration to say that a failure of cybersecurity risk management is actually a failure of executive management.

Finally, although senior executives and boards of directors are not likely to manage cybersecurity themselves, their decisions can have a profound effect on security controls. Structural and functional anomalies have a ripple effect, and because the organizational structure affects nearly every internal and external activity, these effects are replicated across the organization.

9.5 Exception-Based Processes

Exceptions to risk-relevant processes foment security risk management uncertainty and can also create risk factors for information compromise. Exceptions are by definition deviations from standard practice and/or policy, which points to the relevance of policies and standards in reducing cybercomplexity. Moreover, exceptions are more likely to occur if certain structural, functional and/or cultural anomalies exist.

Each exception to an established cybersecurity risk management process and/or to a risk-relevant process, workflow or technology is a potential risk factor for information compromise. The prevalence of exceptions can increase cybercomplexity because of additional and/or varied demands on security risk management. Often these exceptions result in risk-relevant phenomena that occur downstream from the exception. Such exceptions can also add coherently, and thereby also increase cybercomplexity by amplifying the number of risk factors.

In what might seem like a trivial example, exceptions to email address formats or usernames add steps to administrators charged with managing Active Directory and other systems that use email addresses or user names for authentication. The accretion of steps across many systems and processes add risk factors for information compromise that can have a substantive effect on the organization's security risk profile.

The rate of exception occurrence can be modeled mathematically, where exceptions are considered similar to radioactive decay discussed in Chap. 2. Specifically, we assume the generation of exceptions obeys a Poisson process, which expresses the probability a given number of exceptions will occur in a fixed time interval assuming a constant mean rate of occurrence λ and exceptions are independent.

The probability density p(x) of exception occurrences is therefore given by the following expression,

$$p(x) = e^{-\lambda}\lambda^{x}/x! \tag{9.1}$$

For example, suppose λ equals 10 exceptions per process. If 2 processes are included in the analysis, a total of $2 \times 10 = 20$ exceptions would be expected on average. However, suppose we want to know the probability of *exactly* 12 exceptions.

Plugging the numbers into Eq. (9.1), where e is the exponential function equal to 2.72, $\lambda = 10$ and $x = 12$, yields $p(x) = 0.094$ or 9.4%. This result is interesting but we might actually be more interested in the probability *at least* 12 exceptions occur, i.e., the cumulative probability. Determining this probability would entail integrating (9.1) from 12 to infinity assuming the number of exceptions is unlimited. Performing this integration yields a probability of 0.208.

In other words, there is an approximately nine percent probability that exactly 12 exceptions will occur, and a twenty percent probability of 12 exceptions or greater. Of course, the mean exception rate λ would depend on how many processes are included in the analysis since the rate would be expected to vary by process. Therefore, we see the effect of scale in this context. Expression (9.1) and its integral are strongly dependent on λ, and λ depends on the breadth of the IT environment included in the analysis.

Although (9.1) is not a direct measurement of complexity, the resulting calculation could provide insight into risk management uncertainty. Data so derived would represent risk-relevant input to an assessment of security risk management entropy across an IT environment with associated implications to cybercomplexity. Note the perpetual challenge in cybersecurity, indeed in every sub-discipline of security, is to determine precisely (read: quantitatively) how such results correlate with the potential for information compromise.

As noted above, the value of p(x) might only be relevant to a specific process since invoking (9.1) depends upon exceptions occurring at a constant mean rate. Therefore, a generalization of p(x) to the entire IT environment would not be possible if there were high variability in the rate of exceptions across the environment. Of course, such variability might be indicative of other risk-relevant phenomena with their own implications to cybercomplexity.

Exceptions to security risk management processes should be evaluated with respect to their impact on business operations and the potential for information compromise. If these effects are at odds, security governance is required to modulate the security risk continuum calibration point in accordance with the organizational tolerance for risk as depicted in Fig. 9.1. Specifically, a culture that tolerates or encourages exceptions to risk-relevant processes, workflows and technology implementation/configuration should be evaluated in light of its potential effect on cybersecurity risk on an enterprise scale.

9.6 Inconsistent Identity and Access Management

Inconsistent identity and access management is related to the exception-based root cause discussed in the previous section. In fact, the former could be considered a subcategory of the latter. They are treated separately because of the criticality of identity and access management to security risk management. However, the issues cited previously apply equally to this root cause of cybercomplexity.

Access to applications and other IT resources should be driven by an individual's role within an organization. An assigned role is accompanied by access privileges to networked resources. Such privileges are contingent upon satisfying a single criterion: an individual's need-to-know the specific information in question. In other words, an individual network user should only be afforded electronic access to a networked resource if that individual has a need to know the specific information accessible via that resource.

Modern IT environments are replete with applications hosted internally as well as in the Cloud. An individual's "digital identity" is largely defined by the applications and systems to which they have legitimate access privilege, and work-related activities are typically facilitated by one or more of these technologies.

As noted above, access to each application should be dictated by the principle of "need-to-know" since applications are frequently linked to databases containing information that should be restricted to specific network users. In addition, there are presumably a finite number of licenses per application, which demands the number of application users remains below a prescribed limit or additional charges will be incurred.

Figure 9.2 graphically illustrates the high-level architecture of a typical IT environment with respect to both on premise and remotely-hosted applications.

Single sign-on solutions reduce cybercomplexity by streamlining the sign-on process. Streamlining is accomplished by enabling access to all applications to which a user has legitimate access via the use of a single set of log-in credentials (plus a second factor if multi-factor authentication is invoked).

Single sign-on solutions obviate the need for separate credentials per user for each application, thereby simplifying access administration and reducing the number of authentications. In view of the number of applications typically accessible to each network user, and the growing number of Cloud-hosted applications, this process clearly reduces security risk management uncertainty.

A single sign-on solution also decreases the number of risk factors associated with identity authentication, where each authentication represents a risk factor for information compromise. Single sign-on solutions are discussed in Chap. 10 in connection with the effect of standardization.

Administrative details aside, the key point is that an organization's schema to access networked resources should be exclusively determined by an employee's role within the organization and the privileges that accompany that role. Exceptions to the established schema, especially those thwarting automation, will increase complexity and administrative overhead for the reasons cited previously.

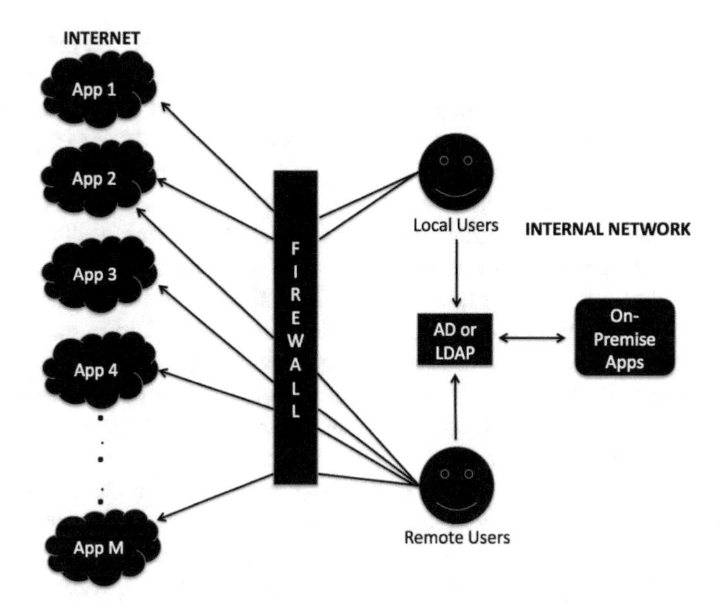

Fig. 9.2 Access to internal and external applications

As with other root causes of cybercomplexity, inconsistent identity and access management can affect other root causes. In this case that other root cause is exception-based processes. Enabling sporadic exceptions might seem relatively harmless, but the exceptions and their effects tend to accumulate and their accretion contributes to security risk management uncertainty.

Furthermore, the potential for unauthorized access to information assets increases as the number of exceptions increases. This effect is due to access privileges becoming stale, waning or ineffective security governance, and the likely growth of the application stack with attendant increases in administrative overhead.

If exceptions are routinely granted by policy, an archeology dig will inevitably ensue months or years later in an attempt to understand why various individuals had been granted access privileges that are inconsistent with their current role. Notably, the contribution to cybercomplexity is clearly scale-dependent since the number of exceptions/risk factors increases with more network users or applications.

9.7 Liberal Internet and Information Access

It is a fact of life that many cyberattacks lure network users to visit malicious websites. Social engineering is rampant, where 22% of all recorded breaches were the result of phishing, and 96% of those breaches occurred via email.[1]

[1] Op. cit. 2020 Verizon Data Breach Investigations Report (DBIR).

Most organizations are keenly aware of this phenomenon, which prompts them to block access to and from malicious websites. In fact, threat awareness and vigilant internet access are two sides of the same security control coin. So-called web filtering means denying network user access to websites that do not align with an organization's tolerance for risk, and arguably is now as essential as a firewall.

Of course organizations can only deny access to sites that are known to be malicious and/or undesirable. Unfortunately such sites are plentiful and easily accessible. Note that certain sites might not comport with an organization's culture and are therefore blocked for reasons other than security. Since the risk posed by such websites is dynamic, web filtering requires an automated solution linked to current threat intelligence and calibrated in accordance with the organizational tolerance for risk.

Organizations can narrow or widen the filter based on that tolerance. Less restrictive filtering will increase cybercomplexity since it results in a larger number of risk factors for information compromise as well as increased uncertainty in security risk management. The down side of increased restrictions is that individuals with a legitimate requirement to access specific content might be denied access to that content. This balance is another example of calibrating the security risk continuum, where the calibration point is set in accordance with the organizational tolerance for risk as depicted in Fig. 9.1.

Internet access is related to the more general issue of information access. Here again cultural issues regarding convenience versus restriction apply. Risk-based restrictions on access to information represent basic security hygiene since the opportunities for information compromise and data leakage increase in proportion to information dissemination.

Once again, such decisions are judgment calls that reflect the organizational culture. In addition, decision consistency and coherence can affect cybercomplexity as discussed in the section on exception-based processes.

Finally, this issue highlights the requirement for centralized security governance. Policies and standards on information access and information technology performance represent a written manifestation of the organizational tolerance for security risk. A security governance process implemented by an executive-level, centralized entity with global authority will enforce compliance on an enterprise scale and thereby promote consistency and coherence across the organization in accordance with its tolerance for risk. Centralized security governance is discussed in Chap. 10 along with other macroscopic security controls.

9.8 Under-Resourced IT Departments

Overly inhibiting the entity with the responsibility for implementing security controls is a root cause of cybersecurity risk with potential repercussions on an enterprise scale. Such entities include the IT Department and/or the cybersecurity risk management function, which is often subsumed within IT Departments.

Lack of support for IT Departments can be rooted in an inherent bias against non-revenue-generating entities. However, this attitude might actually reflect a deeper cultural bias that diminishes the importance of security risk management. For example, it is possible the organization simply values convenience over security restriction.

This situation is not necessarily terrible depending on the issue. However, the attitude needs to be acknowledged by senior executives so that the potential for information compromise is clearly understood. Uncertainty with respect to security risk management could actually increase the potential for information compromise more than liberal security policies in view of the effect of security risk management uncertainty on complexity. In that vein, there must be clarity regarding the tolerance for risk in particular to ensure risk-relevant processes, workflows and technologies are implemented and configured consistently and coherently on an enterprise scale.

The organizational culture is established and sustained by senior executives. Therefore, IT and/or security departments struggling to maintain an appropriate security posture do so because of implicit or explicit mandates from above (or a lack thereof). Moreover, if senior executives do not comply with the cybersecurity policies the organization has published and ostensibly supports, it encourages de facto policy exceptions on an enterprise scale with attendant implications to cybercomplexity.

As always, support for cybersecurity controls is a function of the tolerance for risk and an appreciation for their impact on both security risk management and business operations. The need to strike the proper balance between business facilitation and security restriction demands that individuals formulating and approving security budgets understand the impact of security controls on risk-relevant processes, workflows and technologies as well as the impact of business requirements on the cybersecurity risk profile of the organization.

Chapter 10
Macroscopic Security Controls

10.1 Introduction

Although an assumption of stochastic security risk management is not entirely realistic, it is perhaps equally unrealistic to assume this process is completely deterministic. There is always uncertainty in security risk management, and risk factors for information compromise persist despite the best efforts of competent cybersecurity professionals.

Therefore, complexity is always present in IT environments. The real question is how to manage complexity in this context, which means reducing the potential for information compromise and information-related business disruption on an enterprise scale.

Various complexity metrics were identified in Chap. 8, which represented an outgrowth of a stochastic model of security risk management. These metrics point to the need for an enterprise-level approach to cybersecurity risk management but cannot specify precise adjustments to security controls.

However, this limitation does not invalidate the general lessons that are imparted by these metrics. At a minimum, they comport with intuition regarding the drivers of cybercomplexity as well as provide an objective basis for deploying specific types of security controls, i.e., those with a macroscopic effect.

Security controls that address the root causes of cybersecurity risk would be expected to have such an effect. They would also reduce the likelihood of recurring incidents since root causes are the progenitors of such incidents. The key point is macroscopic security controls must complement tactical efforts at remediation if a comprehensive cybersecurity risk management strategy is the objective.

Interestingly, macroscopic security controls are historically the most difficult to implement. Is this because of their broad impact or is it because the root causes of cybersecurity risk are primal? In other words, do the root causes relate to behaviors and organizational features that are so ingrained that nothing less than systemic changes are required to uproot them?

C. S. Young, *Cybercomplexity*, Advanced Sciences and Technologies for Security Applications, https://doi.org/10.1007/978-3-031-06994-9_10

It is possible both explanations apply. Whatever the reason, the root causes of cybercomplexity tend to promote inconsistency, incoherence, delinquency and/or incomplete security risk management on an enterprise scale. Therefore, the relevant security controls must address the root causes on an equivalent scale if a substantive reduction in the potential for information compromise is the objective.

10.2 Security Acculturation

In Chap. 9 a convenience-driven culture was cited as the most influential driver of cybercomplexity. Its status in this regard in part derives from the fact that other root causes of cybercomplexity relate to and/or evolve from this organizational feature. Therefore it should be no surprise that security acculturation is likely the most influential macroscopic security control.

Security acculturation means creating an organizational culture that includes security in all significant operational decisions. It does not mean that the strictest security measures are necessarily invoked in every instance. In other words, the security implications of business and operational decisions are always considered and actions resulting from these deliberations reflect the organizational tolerance for risk.

However, risk-relevant cultural artifacts are not necessarily communicated to organizational constituents. Furthermore, its precise origins are frequently lost across the vicissitudes of organizational history. As time evolves specific behaviors reflecting the culture are incorporated into day-to-day routines. Sometimes quirky and/or counterproductive cultural artifacts are tolerated, at least until the bottom line is threatened.

For example, it was common practice for individuals to arrive late to meetings at Goldman Sachs. How that practice originated is presumably a mystery but the phenomenon was global and readily acknowledged. One human resources consultant actually referenced this phenomenon in her prepared material.

At other companies staff members might be penalized for continued tardiness. At Goldman Sachs it was tolerated if not tacitly encouraged. Perhaps consistently late arrivals actually decreased effectiveness and/or efficiency. Presumably no one was able to correlate lateness with a negative effect on the bottom line. If such a correlation had been identified, company executives would undoubtedly have moved to change the culture.

Permissiveness with respect to security risk management suggests the tolerance for risk lies closer to business facilitation than security restriction. It is possible that senior executives are ignorant of the likelihood, vulnerability and/or impact of a risk-relevant business practice, which in itself points to organizational culture as a root cause. Alternatively, cybersecurity risk management is considered subordinate to the bottom line, which is yet another relevant cultural artifact.

The culture in any organization is at least tacitly endorsed by the highest levels of management. There are unwritten limits on accepted behavior that are tolerated because they have not crossed some threshold. Lateness to meetings is relatively easy to tolerate, and might even be viewed as an idiosyncratic feature that further distinguishes the company brand.

However, if a significant increase in lateness was perceived to impact the organization's performance it would likely draw senior executive attention. For example, if employees interpreted lateness as a license to skip meetings and the business was impacted as a result, a crackdown would inevitably follow.

Since the breadth and intensity of security controls are linked to the tolerance for risk, and the tolerance for risk reflects the organizational culture, the approach to security risk management is by transitivity a reflection of the organizational culture. The upshot is culture and cybercomplexity are inexorably linked, where the culture either promotes or attenuates risk factor growth and/or risk management uncertainty.

The overarching implication is cybersecurity as practiced on an enterprise scale must be seen to advance the interests of the organization by the organizational leaders if it is to become woven into the cultural fabric of the organization. Absent this type of cultural transformation security risk management efforts will likely be ad hoc and therefore continuously subject to the vagaries inflicted by incoherence and inconsistency.

10.3 Centralized Security Governance

Next to security acculturation centralized security governance is the most significant macroscopic security control. Effective oversight and enforcement of risk-relevant processes facilitates uniformity, which reduces risk management uncertainty on an enterprise scale.

The objective of centralized security governance can be succinctly stated, and was mentioned briefly in Chap. 9, Sect. 6 but bears repeating. Namely, its role is to balance security restrictions with business facilitation in accordance with the organizational tolerance for risk. Centralized security governance ensures security risk management coherence and consistency on an enterprise scale by modulating the relationship between risk-relevant processes, workflows, technologies and security controls. Security control settings ultimately determine the aforementioned balance, which should reflect a juxtaposition of business requirements with the potential for information compromise.

Figure 10.1 graphically illustrates the effect of centralized security governance at a high level.

Centralized security governance is not a single activity. It is a multi-faceted process consisting of both active and passive elements. Active elements are dynamic efforts such as regular engagement by senior executives. Written policies and standards are examples of static elements, where the frequency of policy and standard updates should be dictated by risk-relevant time scales.

Effective security governance requires a written information security policy to establish "ground truth" with respect to risk-relevant behaviors and practices. As mentioned previously, a security policy represents a written manifestation of the

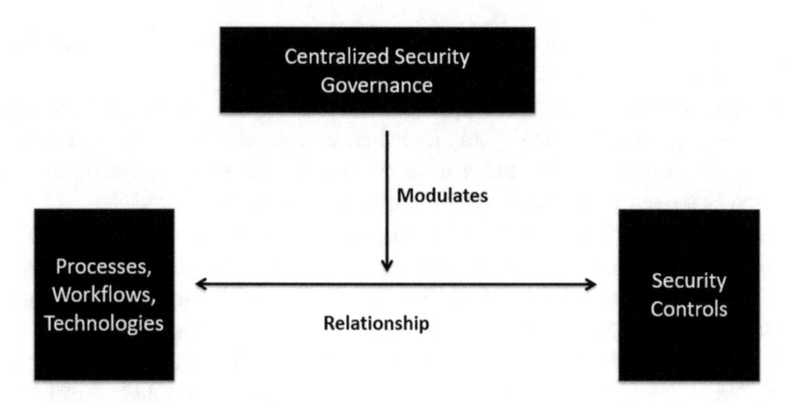

Fig. 10.1 The effect of centralized security governance

organizational tolerance for risk. Since its enforcement ensures the entire organization is in lockstep by promoting uniformity on an enterprise scale, an appropriately structured security policy helps promote risk management consistency and coherence on a similar scale, i.e., the overarching mission of centralized security governance.

The information security policy is the equivalent of the Bible with respect to specifying acceptable information management practices. In fact, a security policy should ideally set forth a set of core principles that are similar to the Ten Commandments in intent if not importance. These principles anchor the various elements of the policy to key cybersecurity themes. All elements of the policy should ultimately derive from or link to these core principles.

Much can be learned about a religion and an organization's security culture from their respective bibles. In fact, a security policy offers exceptional insight into an organization's culture relative to the tolerance for security risk. Importantly, the audience for the information security policy is the community of network users. This document is likely to be overly technical if written by technologists and excessively legalistic if authored by attorneys. Here too a balance must be achieved with respect to readability and the intended message. The policy must be widely read and universally understood if it is to be effective on an enterprise scale and thereby address cybercomplexity.

A security policy is ultimately actioned by network users and system administrators. These actions, or more precisely, the tolerance for their actions, reflects the organizational tolerance for risk. Widespread exceptions and disparities between the policy and individuals' actions are likely indicative of a cultural deficit that must be addressed by senior executives if such disparities are to be broadly remedied.

Of course, no senior executive would be expected to recite the password complexity requirement verbatim. However, such details are established by individuals who ultimately report to these senior executives and who determine the calibration point on the security risk continuum. To reinforce and reiterate the key theme once again, centralized security governance ensures that this calibration point aligns with the organizational tolerance for risk across the enterprise.

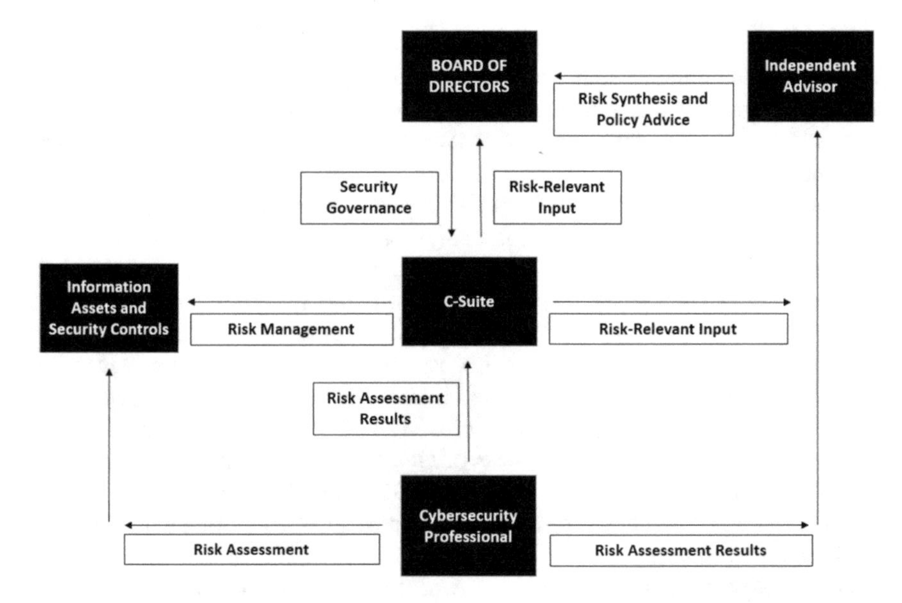

Fig. 10.2 Centralized cybersecurity governance

Effective security governance requires a single entity within the authority makes executive-level decisions about security for the entire organization. An entity with the appropriate authority is necessary to achieve and maintain coherence and consistency with respect to risk-relevant processes, workflows and technologies on an enterprise scale. This unifying force is especially relevant to organizations with multiple venues or business divisions, and particularly if an organization has grown by acquisition.

Figure 10.2 graphically illustrates one version of a centralized cybersecurity governance process.

10.4 Standardization and Compression

This section discusses three processes that have a macroscopic effect on IT environments by reducing the number of risk factors associated with risk-relevant processes, workflows and technologies with attendant consequences to cybercomplexity.

Standardizing risk-relevant processes, workflows and technologies increases IT environment uniformity and therefore decreases unpredictability. The development of formal, i.e., written, standards is intended to standardize IT environments, as the name implies. Standardization within IT environments is most frequently implemented via written standards, which can address a multiplicity of topics.

The motivation for the development of standards is to reduce the potential for information compromise and/or to leverage economies of scale. Technology standards are particularly risk-relevant since they dictate specific system configurations and performance specifications that facilitate administration and/or security requirements.

Compression reduces the number of risk factors by an integer value. The most prominent example of compression is a single sign-on solution. As first mentioned in Chap. 3, a single sign-on solution is an application that enables users to enter their log-in credentials only once in order to access the full spectrum of applications to which they have access privileges. It also facilitates administration of Cloud-based applications, identity and access management and digital identities more generally.

As discussed in Chap. 3, if there are N users and each has access privileges to M applications, each user must authenticate M times to access the full spectrum of applications to which they have access privileges. Therefore, the total number of possible authentications is N × M. Figure 10.3 illustrates the authentication problem for an IT environment consisting of N users and M applications in the absence of a single sign-on solution.

Next consider the same environment where a single sign-on solution has been implemented. In Chap. 3 we observed the total number of authentications depends only on N, the number of network users. The solution compesses M authentications-per-user into a single authentication per-user. Figure 10.4 illustrates the authentication process for an IT environment consisting of N users and M applications using a single sign-on solution.

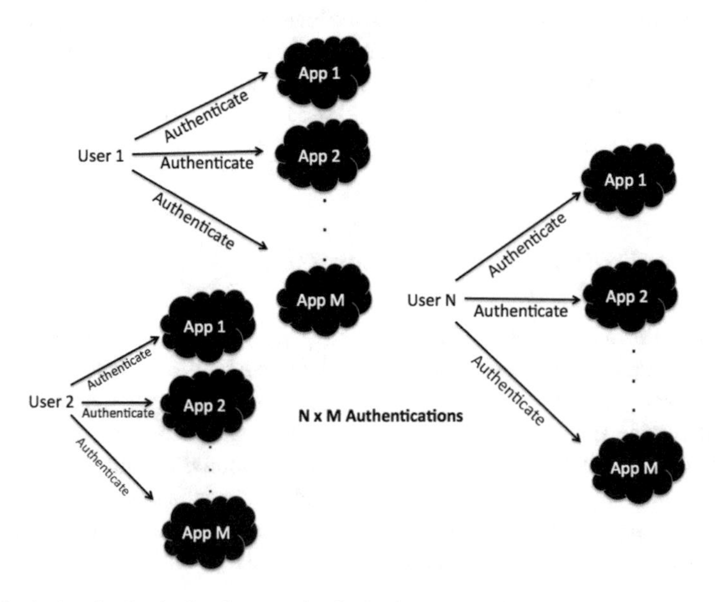

Fig. 10.3 Authentication in the absence of a single sign-on solution

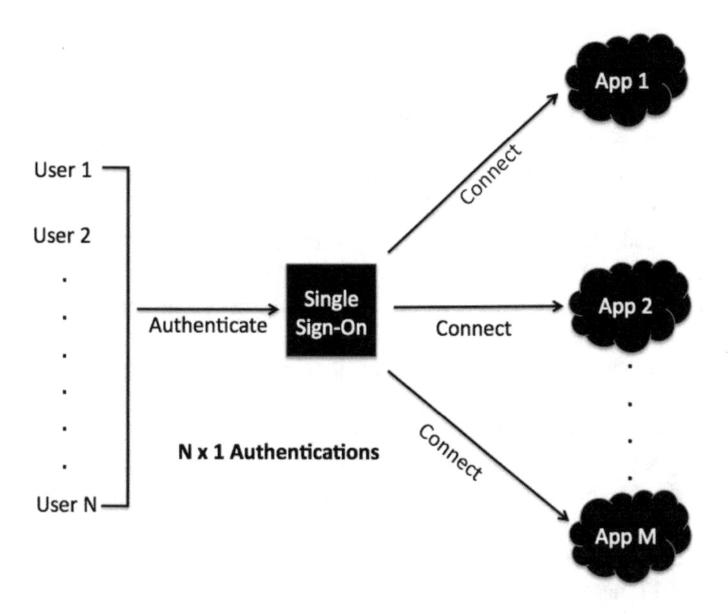

Fig. 10.4 Authentication using a single sign-on solution

However, note that a single sign-on solution also concentrates risk. Specifically, it increases the impact component of risk since the compromise of a single resource could simultaneously facilitate unauthorized access to M applications. In the words of the American writer, Mark Twain, "If you put your eggs in one basket, you had better watch that basket!".

As noted in the discussion on relative complexity in Chap. 8, reducing complexity does not necessarily eliminate the potential for information compromise. There are numerous ways the potential for information compromise can be affected that have nothing to do with complexity. In that vein, it is possible that reducing cybercomplexity actually increases the potential for information compromise.

Consider the unlikely scenario where an organization implements two-character passwords in order to appease network users by massively increasing convenience. Cybercomplexity has been decreased by increasing uniformity but the potential for information compromise has certainly increased. The lesson is it is imperative to consider all effects of a security control before implementation.

10.5 Role-Based Identity and Access Management

Identity and access management (IAM) is the schema or process that governs access to information resources. Specifically, IAM specifies "who" has access to "what." The risk-relevance of IAM cannot be overstated.

On what basis should assignments of IT resources be made? A logical, i.e., risk-based, approach would be to assign IT resources according to an individual's role within the organization. Risk management would automatically be incorporated into such assignments since access privileges should be dictated by a need to know the particular information relevant to a specific role.

Note certain individuals could have multiple roles requiring different access privileges. Other individuals might have intersecting roles as defined by their position, function and/or location, wherein access privileges might actually be in conflict. The upshot is effective implementation of IAM is not always straightforward but it is certainly always necessary.

Establishing the criteria that define each role and identifying the hierarchy of those criteria are critical elements of IAM. Clearly the requirement to authenticate the identity of individuals with access privileges is partly driven by the likelihood, vulnerability and/or impact of a compromise of the information being protected.

The nature of networking is such that individuals must communicate with individuals who can neither be seen nor heard, and therefore must be authenticated without the benefit of sensory input. Accessing and sharing information assets are the *raisons d'être* of an IT network, and therefore the authentication process is continuously repeated. Moreover, in the absence of sensory-based authentication, relevant security controls must establish *trust* in the identity of network users via other mechanisms.

Although quantitative thresholds for trust cannot be established, qualitative thresholds are implicit in certain security controls, e.g., multi-factor authentication. Requirements for establishing trust in a network user's identity deserve special attention because of their centrality in facilitating access to information assets, and are therefore discussed in more detail in Chap. 11.

10.6 Education, Training and Threat Awareness

The interaction of humans and technology is a key feature of IT environments. This relationship is not simply synergistic, and is probably better described as co-dependent. A computer network is not much use without users, where users require computers, software and a networking infrastructure to effectively communicate.

IT environments facilitate the interaction between humans and machines. This condition is simultaneously a crowning achievement and an ongoing nightmare since computers have simultaneously revolutionized information sharing and exponentially enhanced the potential for information compromise.

Because the interplay between humans and machines is fundamental to IT environments it is also fundamental to cybersecurity risk management. This interplay partly explains why security governance is so significant. Cybersecurity controls are often technology-based, and these controls are inextricably linked to risk-relevant business processes, workflows and technologies that facilitate information sharing by network users across the enterprise.

The absence of centralized security governance will likely introduce variability within IT environments that results in security risk management uncertainty as well as an increase in the number risk factors for information compromise on an enterprise scale. This condition is especially true for IT environments with numerous network users who (1) access disparate applications, (2) reside in diverse regions and/or (3) utilize a spectrum of technologies. Each of these features applies to modern organizations.

A substantive reduction in cybercomplexity will require modifications to computing behavior. Therefore, security governance, like other macroscopic security controls, is necessary but not sufficient to address the effects of cybercomplexity.

For example, merely publishing a security policy is not sufficient to address certain threats if network users are unaware of those threats. Cybersecurity education, training and threat awareness on an enterprise scale is also required in order to have a meaningful effect. These programs can take many forms, but experience suggests messages on acceptable computing behavior must be repeated if their content is to be assimilated by the majority of network users.

Note education, training and threat awareness can be an effective security control if implemented properly, but quantifying its effect is problematic. For example, phishing tests are frequently used to measure network user susceptibility to social engineering and the salutary effects of a security training program. The general rule of thumb based on anecdote is that approximately thirty-three percent of network users click on bad links even *after* training. Recall this scenario was discussed in conjunction with the relative complexity metric in Chap. 8.

Since a phishing attack only requires a single errant email response, and the effect of training appears to have a *de facto* lower bound, it is not clear the thirty-three percent figure is a particularly useful risk metric. It seems more prudent to assume one or more network users will ultimately fall victim to social engineering and to deploy security controls to address that eventuality.

Constraining on-line behavior, i.e., reducing permissiveness, will generally reduce IT environment variability and therefore decrease security risk management uncertainty with an accompanying impact on cybercomplexity. Achieving consistency and coherence in user behavior on an enterprise scale, a *sine qua non* for reducing cybercomplexity, requires regular messaging to the networked community. This approach to communicating increases the likelihood users will comply with policies, which translates to a more complete and consistent alignment with the organizational tolerance for security risk. It also increases general awareness of any immediate threats to the IT environment.

10.7 Internet Intelligence

The rapid rise in the popularity of the internet has been breathtaking as noted in the book's Introduction. The extreme dependence on its continued functionality is equally astonishing. Operations would literally cease at most organizations if internet access were significantly disrupted for a protracted period. Households would be equally discombobulated since people conduct much of their personal business on-line.

Since the internet is simultaneously a vehicle for Good and Evil, it is imperative to know thy enemy. At a minimum such advice means being aware of, and proactively intervening in, attempts to connect to malicious web sites. However, because such sites are numerous and change with time, threat awareness and pro-active intervention must be continuous. In addition, they require automation since threat awareness must trigger intervention in real time.

A practical cybersecurity risk management strategy must include security controls such as blocking outbound access via a firewall or enabling Domain Name Resolution filtering to prevent connecting to malicious web sites from internal IT resources. Threat awareness and training of network users with respect to safe on-line behavior will further decrease the exposure to malware, *et al.* per the previous section. In fact, the Internet Intelligence macroscopic security control complements the Education, Training and Threat Awareness macroscopic security control discussed previously.

The frequency of attacks via unauthorized network access suggests that implementing automated web filtering to prevent such occurrences is on par with multi-factor authentication in terms of strategic and tactical significance. Complexity reduction is a significant by-product of this security control since narrowing the spectrum of accessible websites simultaneously reduces risk management uncertainty and the number of risk factors for information compromise.

10.8 Data and Resource Minimization

The intent of data and resource minimization is to reduce the IT environment attack surface, which is defined as the breadth of information assets susceptible to attack by an adversary. To be clear, data and resource minimization means removing networked resources from IT environments that are not required to facilitate business operations, fulfill a legal/regulatory requirement or ensure effective disaster recovery.

Data "sprawl" is an increasingly common phenomenon in modern IT environments. Data of varying types and confidentiality are routinely spread across internally and externally-managed resources. In addition, the proliferation of various sites requiring access credentials increases opportunities for leaked credentials. Since network users frequently reuse on-line credentials, one compromised site can result in unauthorized access to other sites, thereby increasing the potential for information compromise.

Data sprawl has been amplified by the proliferation of Cloud resources to store data and transfer files. Internal storage solutions add to the number of storage options. Limiting such options decreases cybercomplexity and reduces the potential for information compromise via data leakage.

In view of the spectrum of options and number of customers seeking to leverage these options, a holistic data storage and transfer strategy is can be non-trivial to develop and especially challenging to enforce. This condition provides additional justification for implementing another macroscopic security control, centralized security governance. The security governance process should include an organizational policy that aligns storage diversity and business requirements with the tolerance for risk as manifest in the relevant security controls, e.g., proper authentication and limits on storage duration.

Reducing the number of supported resources will reduce cybercomplexity in a manner similar to data sprawl. Risk factors for information compromise accompany almost any networked resource. Software applications require continuous updating via automated installation of security patches, and each vulnerability represents at least one risk factor for information compromise.

Therefore, the number of applications included in the canonical image for endpoints should be only those required to satisfy business requirements. Note the magnitude of this effect is scale-dependent. Culling applications replicated across the enterprise could potentially eliminate thousands of vulnerabilities depending on the number of endpoints and the rate at which vulnerabilities are addressed (or not).

Although it is not equivalent to decommissioning hardware, reducing the number of supported applications or limiting data sprawl, the migration of internal IT resources to a full-cloud solution or Software as a Service (SaaS) has the added advantage of transferring security risk management responsibilities to a third party. SaaS solutions also reduce the internal IT footprint, noting there are both hard and soft costs associated with actively managing IT resources. Deployment of SaaS applications is frequently motivated by cost but it also reduces cybercomplexity.

The effect on cybercomplexity can be exponential if resource decommissioning, application reduction and migration to SaaS applications are repeated across an IT environment. As always, decisions on data and resource minimization must be balanced against the business case for continuing support.

In summary, macroscopic security controls represent a key element of a cybersecurity risk management strategy. In general these controls are agnostic to specific technologies since they affect the approach to security risk management. Ultimately they increase uniformity across the enterprise, which reduces security risk management uncertainty. They can also reduce the number of risk factors by eliminating elements from an IT environment. In both instances the net effect is to reduce the potential for information compromise on an enterprise scale, and thereby reduce the magnitude of cybercomplexity.

Chapter 11
Trust and Identity Authentication

11.1 Introduction

Common sense dictates that numerous, largely anonymous network users interacting with disparate technologies pursuant to sharing information would be accompanied by an increased potential for information compromise. That said, other technological environments that qualify as complex operate without an issue. Either complexity alone does not explain the rich history of information compromises or IT environments are somehow unique. The evidence suggests both explanations apply.

In fact, there are many examples of systems that have successfully addressed complexity as a risk factor for faulty performance.

Undoubtedly many of those systems contain interdependent sub-systems and in that respect are similar to IT environments. Therefore, system interdependence and the sheer number of elements would not by themselves explain the history of information compromises.

Consider the breadth of instruments and technologies used in modern commercial aircraft. One would expect complexity would plague the airline industry and would be reflected in safety statistics. However, the salient point is only qualified pilots interact with those instruments. Imagine a scenario where each passenger could influence the plane's performance while in flight. Clearly the safety of all aboard would be much less assured.

The contention is the ongoing interaction between network users and information technologies distinguishes IT environments from other complex, technology-rich environments. Compounding the problem is network users have significant independence to perform a host of functions that bring them into electronic contact with the outside world. Notable among these functions is emailing and internet access.

Furthermore, the behavior of a single network user affects all networks, applications and users with whom he or she connects or interacts. The interactions between network users and technologies affect cybercomplexity in two ways.

First, they clearly increase risk management uncertainty on an enterprise scale, where the magnitude of uncertainty is fluctuating in time. At any moment any network

C. S. Young, *Cybercomplexity*, Advanced Sciences and Technologies for Security Applications, https://doi.org/10.1007/978-3-031-06994-9_11

user could be connecting to or enabling a connection to a malicious third party. Second, risk factors associated with on-line activities such as identity authentication, authorizing access to information assets, asset provisioning/de-provisioning, etc. are regularly if intermittently surfacing.

In the absence of these interactions IT environments would be constrained to function in a more prescribed and therefore predictable manner. Consequently, the increase in cybercomplexity that accompanies certain risk-relevant activities would be reduced. Aside from technology malfunctions, where the average rate can be predicted using statistical methods, the technology would essentially do its thing with reasonable predictability if maintained properly.[1]

Of course an IT environment without network users is illusory if not an outright contradiction. The purpose of an IT environment is to enable information access and sharing. Therefore, the answer to more effective assessments of cybersecurity risk is not to reflexively make wholesale changes to IT environments, but rather we should initially seek to modify such assessments and adjust security risk management according to the results.

Certain issues disproportionately contribute to the potential for information compromise and therefore are especially risk-relevant. One such issue is uncertainty in the identity of individuals requesting access to information assets. To that end, the principal objective of authentication as a security control is to establish *trust* in the identity of each network user accessing an information asset each time access is attempted.

Establishing a quantitative threshold for trust would be ideal. However, in the absence of a properly calibrated "trust meter," authentication processes must rely on qualitative methods that are known to decrease the likelihood of unauthorized access. Such methods prominently include multi-factor authentication and correlation.

Although their precise effect on the potential for information compromise cannot be quantified, these methods have a quantitative-like rationale. For example, each additional factor used in multi-factor authentication exponentially decreases the likelihood of unauthorized access. This statement is easily proven but cannot be precisely calculated as demonstrated immediately below.

Let p_1 equal the probability one authentication factor is compromised and p_2 equal the probability a second, independent authentication factor is compromised. Trivially, the probability both risk factors are compromised equals p_1 times p_2, which will always be less than either p_1 or p_2 since probabilities are less than one by definition. Therefore, the overall probability decreases exponentially with an increasing number of authentication factors.

However, the precise reduction in the potential for unauthorized access to information assets cannot be calculated since it is impossible to determine either p_1 or p_2. It follows immediately that the number of required authentication factors also cannot be calculated. The usual number of factors used in operational scenarios is two.

[1] Weibull probability distributions are used to describe observed failures. They are widely used in reliability and survival analysis.

This number should be driven by the organizational tolerance for risk in conjunction with the operational and financial burdens associated with implementation. More factors require more resources. However, the tolerance can be equally difficult to ascertain. Ultimately, a qualitative threshold for trust must be established where decisions on identity authentication requirements are inherently qualitative but still require rigor.

Finally, the connection between trust and identity uncertainty suggests the potential for a stochastic characterization of identity management. Although a probabilistic approach will not lead to the development of a "trust meter," it can facilitate a general framework for trust with respect to identity and thereby identify classes of scenarios requiring enhanced authentication.

11.2 The Fundamentals of Trust

A repeated theme within the cybersecurity profession is the concept of "zero trust." Zero trust means no system or information technology can be accessed without proper identity authentication. Trust in this context is defined as sufficient confidence in the identity of an individual seeking access to a particular information asset. Clearly there is a lot riding on the word "sufficient."

It is necessary to understand the relevance of trust as a security control, and in particular why it is fundamental to identity authentication. The requirement for trust in this context derives from the fact that interactions with information technologies are initiated by humans whose identity cannot be directly verified by human perception. Establishing sufficient confidence in the identity of network users is a prerequisite for confirming the integrity of online activity.

Humans can perform identity authentication in person simply by invoking their sensory perception in conjunction with exceptional powers of recognition. Sensory input coupled with sophisticated neurological processing enables humans to distinguish between thousands and possibly millions of distinct individuals. It is unclear if there is a limit to this capability in practice.[2]

Such capabilities are generally taken for granted but are noticeably absent when sensory perception is not an option. Consider an individual requesting to borrow a cup of sugar from a neighbor. The borrower is permitted entry to the neighbor's home because visual, aural and/or olfactory clues are used to confirm his or her identity. This process of authentication is nearly instantaneous, where presumably the skill has been honed by natural selection over millennia.

There is typically little uncertainty in identity authentication via visual recognition, especially if there is a history of interactions. Automated facial recognition can now mimic human capabilities. The accuracy of such systems can be quantitatively

[2] Humans can reportedly detect several million distinct colors, half a million musical tones and over a trillion scents! https://www.healthline.com/health-news/strange-human-noses-can-detect-more-than-a-trillion-scents-032014.

assessed via the rates of true and false positives and negatives. But it is not clear if these systems can yet approach human capabilities especially in low-light conditions and/or if the image is otherwise sub-optimal. Humans are typically not affected except under extreme conditions.

Networked systems must compensate for the absence of visual and aural clues using protocols specifically designed to reduce identity uncertainty. As noted in the discussion there is an implicit yet qualitative threshold for achieving sufficient trust in identity authentication.

Trust between humans more generally reflects a combination of duration and history. The working assumption in establishing trust is that past is prologue. For example, people trust their parents, spouse, best friend, dog, etc., because of a history of interactions that generates confidence in future interactions. Individuals with whom a history of favorable interactions does not exist require additional mechanisms to establish a threshold for trust.

Note that trust is essential to human interaction and a functioning society. Imagine if the viability of produce, the safety of commercial aircraft, the privacy of communications, the electrical grounding of electronic devices, the competence of physicians, and the trustworthiness of family and friends were constantly in question.

A functioning society requires trust in both people and processes. Humans employ a conscious if informal vetting system to establish trust. As noted above, we trust people with whom we have a longstanding and historically favorable relationship. In contrast, one is more circumspect when interacting with individuals with whom there is less familiarity or a questionable history until some subjective threshold is achieved. Physical and virtual forms of identity authentication are equivalent in their respective objectives if not the precise methodologies used to achieve those objectives.

Continued trust is affected by the most recent history of interactions. For example, trust in the neighbor requesting to borrow a cup of sugar continues unabated if and only if that neighbor's current and past behavior correlate, which is a fundamental criterion for achieving trustworthy status.

The criteria for establishing trust are generally different for a relationship and a process. Experience with a given process increases confidence in its effectiveness. However, for certain processes, independent arbiters possessing specific expertise are required to establish and maintain trust.

For example, the United States government relies on experts at the National Institute of Standards and Technology (NIST) to evaluate the quality of commercial encryption algorithms. Similarly, the Federal Drug Administration (FDA) evaluates the efficacy of drugs, the Center for Disease Control (CDC) provides risk-relevant information on diseases and the Federal Aviation Administration (FAA) evaluates aircraft safety. One essential role of government is to function as a source of public trust, which explains the furor when that trust is violated.

Each session on the internet establishes an ephemeral relationship between online entities. Each session, and therefore each relationship, is unique since communication pathways and/or network communicators are not static from session-to-session. IP-based technologies facilitate communication on a fantastic scale, but they also introduce uncertainty that vitiates trust in identity. The inherent uncertainty that accompanies ephemeral relationships must be addressed using specific security controls.

Information can only be freely exchanged when certain security controls sufficiently increase the likelihood network communicators are legitimate. The identity of network communicators must be authenticated each time access to information assets is requested in keeping with the principle of zero trust.

In the absence of a quantifiable threshold for trust there must be qualitative standards that dictate the acceptable limits on risk for each on-line session. To that end, combinations of security controls or specific identifiers are used to establish trust in identity authentication.

Clearly trust is contextual. It is affected by features of the IT environment including the scale of the particular environment in question. Consider the thought experiment conducted in Chap. 7 consisting of a minimalist networking setup. In that scenario the opportunities for both information compromise and information transmission were effectively nil. The communication constraints promoted simplification, which eliminated numerous vectors for information loss or leakage.

However, simplification in that context was accompanied by significant restrictions on communication. In that idealized IT environment, trust in the identity of network users and the integrity of communications would be relatively easy to achieve because of the limited scale of the IT environment and the absence of spatiotemporal variability.

In contrast, typical commercial IT environments might have thousands of network users logging on and off throughout the day from disparate locations beyond the perimeter firewall. These environments could also contain hundreds of switches and routers in addition to thousands of endpoints, e.g., desktops and mobile devices. Both wired and wireless devices would likely be used to transmit messages and send documents via the network. Thousands of internet connections might be initiated and discontinued per-hour, where the relevant packets traverse routes that are invisible to communicators and vary with each connection.

Trust in the identity of entities connecting to a typical IT environment is reduced relative to the simple local area network described above. An increase in the potential for information compromise arises as a result of the myriad of fluctuations that occur on a macroscopic scale thereby increasing security risk management uncertainty across the enterprise.

As a result, additional security controls are required to increase the likelihood network users are legitimate. Note as the assessment scale is reduced individual elements within the IT environment would become more in focus at the expense of the big picture. One might be tempted to assume security controls should increase without limit and thereby establish an exceedingly high threshold for trust. On its face such a strategy might greatly reduce the potential for information compromise. However, this strategy would come at a price. Perhaps ironically, that price is a reduction in trust in the authentication process itself.

Once again consider the use of multi-factor authentication, traditionally implemented using "something-you-possess" in combination with "something-you-know." As alluded to previously, two factors are generally accepted as the industry standard. Initial resistance to multi-factor authentication has gradually evaporated due its widespread use and the unambiguous protection it affords against social engineering, a common cybersecurity attack vector.

Successful identity authentication using both factors translates into lower identity uncertainty. The security conferred by two factors is generally thought to be good enough assuming the channels used to transmit each factor are not themselves compromised. Identity uncertainty would be expected to decrease exponentially with each additional factor as also discussed previously.

However, the resulting decrease in the potential for information compromise would likely be offset by the diminishing returns that frequently accompany inconvenience, especially if the inconvenience is believed to be without purpose. Ultimately, workarounds will be sought to circumvent security controls that are perceived to be disproportionate to the magnitude of risk. In fact, a security control can itself become a risk factor for information compromise if it is viewed as overly burdensome and/or superfluous.

We can speculate regarding how the threshold for achieving trust in identity authentication is actually achieved. For example, the magnitude of trust could be depicted as a smooth function normalized to one in Fig. 11.1. Trust asymptotically approaches a maximum as the modes of authentication increase. In other words complete trust in identity approaches a maximum value but never actually achieves this theoretical limit.

However, it is also possible trust in identity authentication is actually a stepwise function as shown in Fig. 11.2, where each additional mode results in a discrete uptick in trust. Again the trust in identity authentication asymptotically approaches a maximum value.

Fig. 11.1 Smooth increase in trust in identity authentication

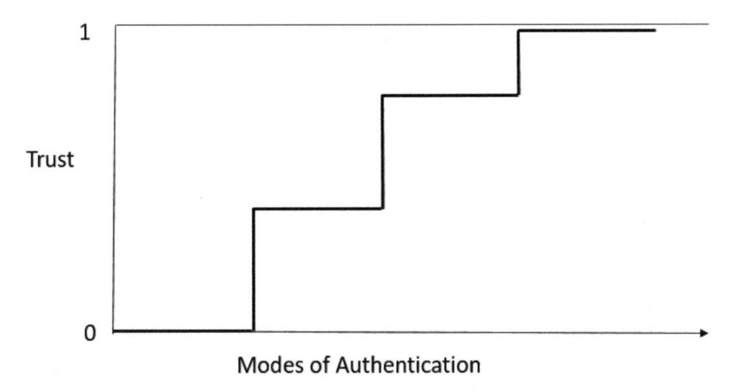

Fig. 11.2 Stepwise increase in trust in identity authentication

The astute reader will appreciate that neither Fig. 11.1 nor Fig. 11.2 provides much useful operational guidance. Although each representation might be directionally correct we are still challenged to calibrate the modes of authentication. Therefore, it is impossible to determine which figure represents a more accurate depiction of reality.

Identity uncertainty must be inversely related to trust in identity authentication. Increasing trust decreases uncertainty by definition, and this general statement also applies to identity authentication. Furthermore, identity authentication must itself be inversely related to the number of authentication factors. Of course, the use of two authentication factors is stronger than one, three factors is stronger than two, etc. "Stronger" in this context means identity uncertainty decreases with an increasing number of authenticating factors.

Trust in identity authentication is therefore trivially seen to be directly proportional to the number of factors used for authentication. Unfortunately, this statement has also not significantly progressed our thinking since it has not revealed the precise effect of each authentication factor. Although expression (3) is nominally quantitative, a *bona fide* prescription for security risk management would specify the exact number of authentication factors required to achieve *sufficient* trust in identity.

Figure 11.3 is a high-level characterization of the relationship between cyber-complexity and trust in identity authentication. It also indicates the role of scale and scaling relations in assessing complexity within IT environments. In the next section a stochastic approach to identity authentication will be discussed. The objective is to specify categories of trust scenarios. That discussion will also reveal the inherent difficulties that accompany any stochastic formulation of security risk management.

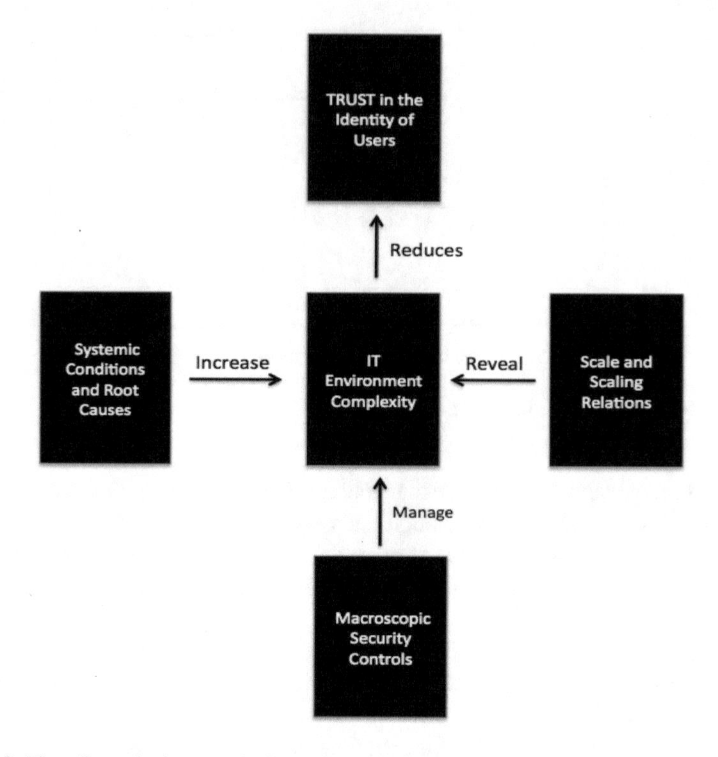

Fig. 11.3 The effect of cybercomplexity on trust in identity authentication

11.3 Identity Authentication Entropy and Trust

We assume trust and the absence of trust can once again be viewed as two outcomes of a binary stochastic risk management process, i.e., "trust management." To be clear, trust management is actually trust in identity authentication. The entropy of the probability distribution for trust management would be identical to Fig. 6.9 in keeping with any binary stochastic process.

Trust management can now be represented as a continuum of values. The two extremes are complete trust uncertainty, i.e., the probability of identity authentication equals 0.5 and therefore the information entropy must equal 1, and complete trust certainty, i.e., the probability of identity authentication equals 1 and therefore the information entropy must equal zero.

Figure 11.4 illustrates the trust management continuum, which is analogous in form to the security risk continuum illustrated in Fig. 9.1.

Stochastic formulations of both cybercomplexity and trust management are rooted in the inherent uncertainty/diversity of probability distributions. Cybercomplexity results from the uncertainty/diversity of the security risk management process/message source applied to an IT environment or subset of an environment.

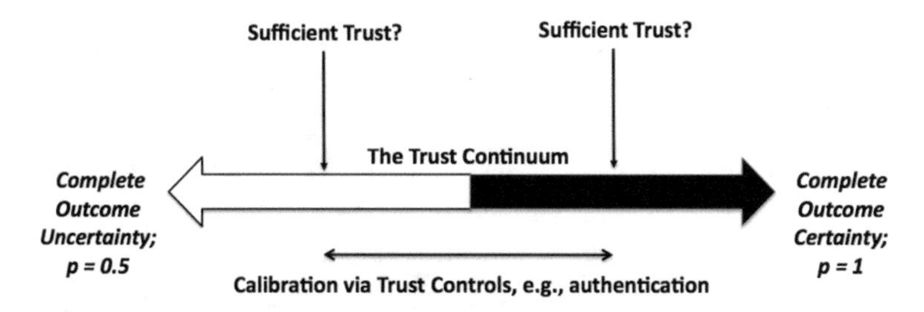

Fig. 11.4 Trust management continuum

However, identity authentication uncertainty is only one of many potential sources of uncertainty affecting the potential for information compromise on an enterprise scale. In other words, uncertainty in identity authentication is one contributor to security risk management entropy in an IT environment.

If trust management is assumed to be a binary stochastic process as described above, the number of probable states would be calculated from the entropy (H) of the trust management process in combination with the number of authentications (M). The number of probable states (T) would again be given by 2^{MH}.

For example, if a thousand authentications were attempted and the entropy of the authentication process is estimated to be 0.1, there are $T = 2^{MH} = 2^{1000 \text{ x } 0.1} = 2^{100}$ probable states of trust. The probability of a specific probable state is $2^{-MH} = 1/2^{100}$. Given the numerous authentications performed in a commercial IT environment, T will always be a large number unless H is close to zero. Note in the limit, i.e., only two possible network communicators, the requirement for trust is eliminated since H equals zero and T becomes unity.

Candidly, the operational value of this absolute metric is not immediately apparent. However, another metric might be more informative.

Recall the relative complexity metric C_r also described in Chap. 8. Because the identity authentication uncertainty must be extremely small in an operationally viable IT environment, we now compare the deviation with respect to the *minimum* value of T, i.e., when H equals zero or T equals one. In other words, the metric for relative trust would measure the deviation from a minimum identity authentication uncertainty condition.

Specifically, a metric for *relative trust* T_r might be $2^0 - 2^{-MH} = 1 - 2^{-MH}$. If the identity authentication entropy H for an IT environment equals zero, the relative trust metric equals zero, which is ideal from a trust management perspective. However, if the product of M and H is non-zero, the value of T_r approaches 1 as M and/or H increases. Therefore, the closer T_r is to unity, the greater the deviation from an IT environment where identity authentication is a certainty.

Of course, due to the symmetry of binary stochastic processes such a condition includes IT environments where all identity authentications are unauthenticated. In addition, if a linear representation of T_r is chosen, H would need to be scaled appropriately since 2^{-MH} will rapidly become infinitesimally small otherwise. A logarithmic representation of T_r would be linear in M and H.

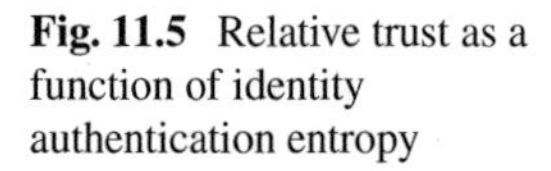
Fig. 11.5 Relative trust as a function of identity authentication entropy

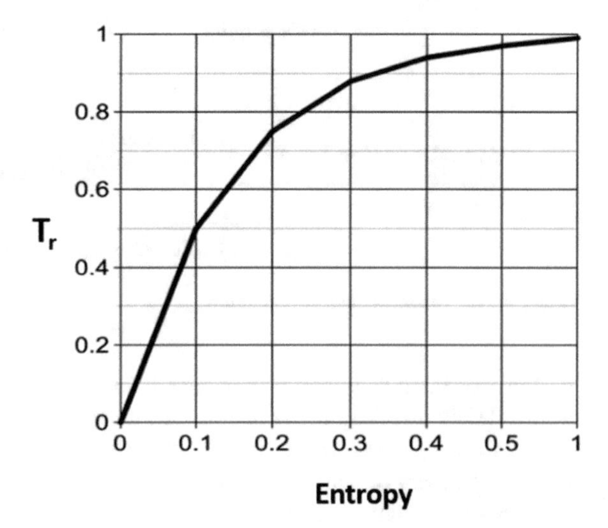

Figure 11.5 plots relative trust versus entropy for an IT environment with 10 authentications.

Figures 11.6 and 11.7 are two hypothesized representations of the magnitude of trust in identity authentication as a function of identity authentication entropy. The entropy changes as the confidence in authentication is incrementally reduced. For example, the confidence might be proportional to the number of authentication factors. A decrease in the number of authentication factors will increase the entropy. These figures are merely inverse representations of Figs. 11.1 and 11.2.

The above approach could never be considered for authenticating the identity of single individuals. To be clear, that is not its purpose. A stochastic formulation of a security risk management process is mostly useful in understanding the drivers of a risk-relevant condition. Furthermore, trust in this context is a statistical phenomenon

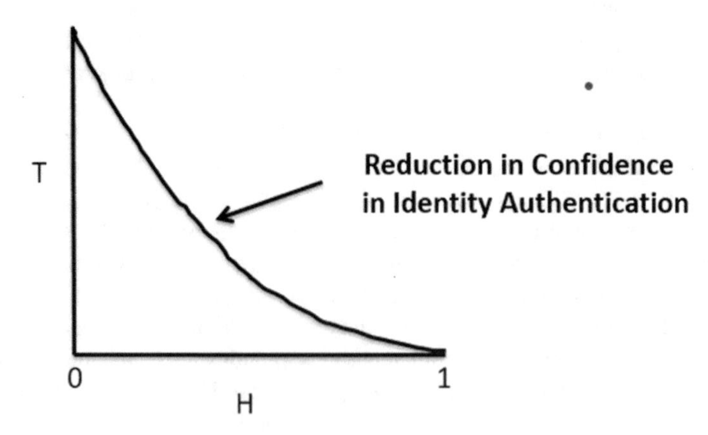

Fig. 11.6 Trust as a function of identity authentication entropy (1)

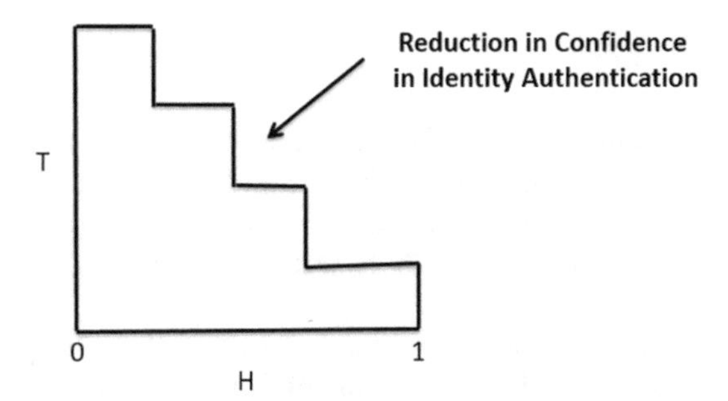

Fig. 11.7 Trust as a function of identity authentication entropy (2)

and therefore is inherently macroscopic. It is neither meaningful nor accurate to apply this process to individuals for the purpose of identity authentication.

The importance of reducing identity authentication uncertainty seems irrefutable. Realistically, this model merely confirms what seems obvious and mostly provides an analytic validation of intuition.

Figure 11.8 plots the entropy of a binary stochastic trust management process. As expected, the entropy/uncertainty in identity authentication increases linearly with the log (base 2) of the probability of a correct identification. The probability is inversely related to the number of identities, so $p = 0.5$ corresponds to two possible identities, $p = 0.25$ corresponds to 4 identities, etc.

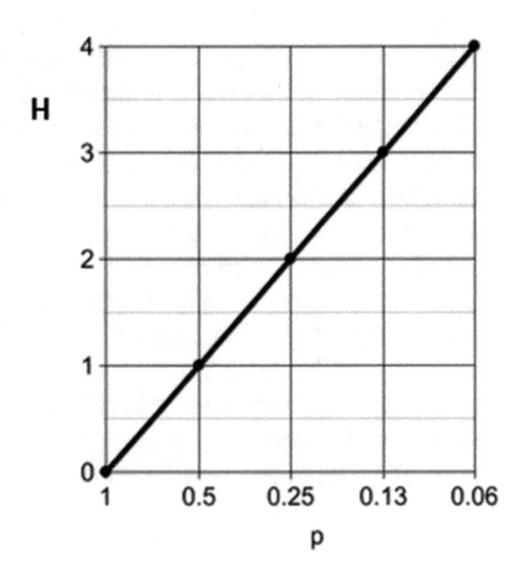

Fig. 11.8 Identity uncertainty in trust management

11.4 Correlation and Trust

Achieving trust in identity authentication is more difficult in the presence of cybercomplexity. The number of risk-relevant scenarios that accompanies each additional network user requesting access to information assets exponentially increases the number of risk factors for information compromise as well as contributes to security risk management uncertainty.

The method of correlation identifies *probable* scenarios based on historical precedent, and has the overarching effect of reducing cybercomplexity. Historical scenarios bias the identity authentication process and suggest a particular scenario has crossed the minimum trust threshold. Specifically, trust in the identity of a network communicator increases if that communicator conforms to a recognized pattern of behavior and is therefore considered non-threatening based on historical precedent.

To reiterate, the objective of trust management is to reduce uncertainty in identity authentication. A sufficient reduction in uncertainty is achieved through the confirmation of specific features or attributes historically associated with the individual requesting access. These features include previous devices, networks and physical locations.

The fundamental presumption is past is prologue. Specifically, a previously trusted individual possessing a specific set of attributes who was granted system access at time t is likely to be the same individual requesting system access at time $t + \Delta t$ if this set of attributes is recognized. This method of establishing trust in identity authentication is known as *correlation.*

For example, if the individual requesting electronic access is initiating the request from a previously recognized country, device and network, the presumption is the required threshold for identity authentication has been achieved.

The fact that these attributes are not present is not necessarily indicative of malicious behavior. It is possible that a legitimate requestor is using a new device, is located in a new country and/or is launching the request from a new network. But the minimum if qualitative threshold has not been crossed for identity authentication noting the precise reduction in risk of unauthorized electronic access as a function of each feature type cannot be determined.

Quantitative correlation metrics are possible if the trends in two variables can be compared point-by-point. There are several types of so-called correlation coefficients that can yield risk-relevant metrics if the data are available to perform the required calculations. Perhaps the most common form of correlation metric is the Pearson Product Moment Correlation (PPMC), which quantifies the linearity of a relationship between two variables.

PPMC values range between minus one and positive one. A plus one (+1) coefficient means there is perfect correlation between two variables. In other words, an increase in one variable is accompanied by a proportionate increase in the other variable, which the reader will recognize as the defining characteristic of linearity. Perfect anti-correlation implies an increase in one variable results in a similar decrease in the other variable. Zero correlation means a linear relationship between the two variables does not exist.

Figure 11.9 depicts two perfectly correlated variables. In this case the PPMC equals plus one since each successive value along the horizontal axis results in an increase of two units along the vertical axis for both trend lines. In other words, the two trends have identical slopes so there is a perfect linear correlation.

Figure 11.10 graphs two variables with a PPMC of minus one (−1), i.e., perfect anti-correlation. In this example the x-axis is time (year), and the y-axis is both the number of threat incidents and the annual budget in arbitrary units. If an increase in the annual number of threat incidents is perfectly anti-correlated with the annual security budget it means security incidents are increasing at the same rate the security

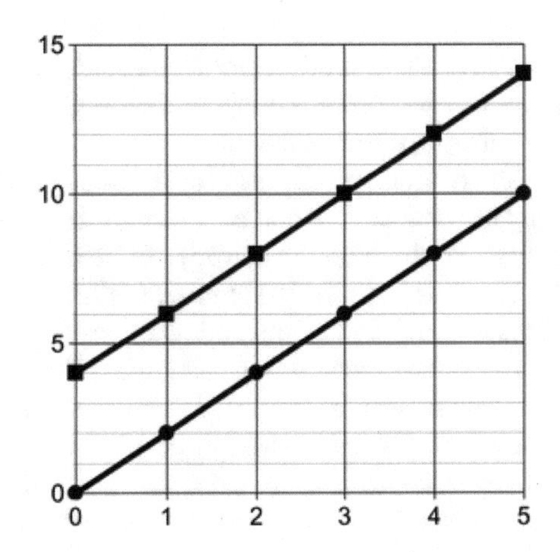

Fig. 11.9 Perfectly correlated variables

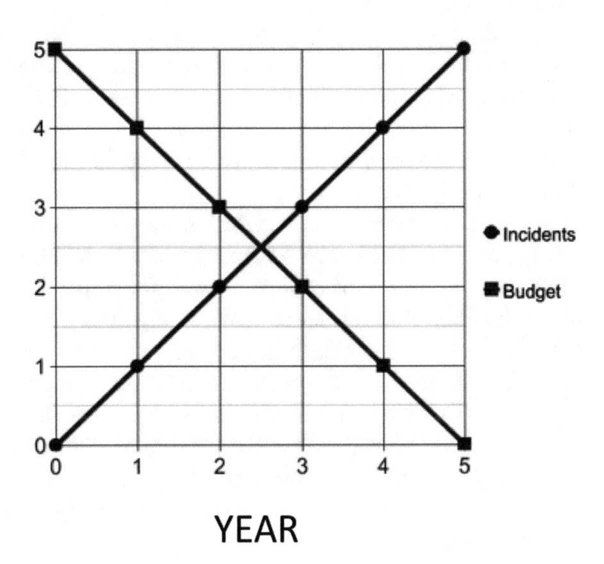

Fig. 11.10 Perfectly anti-correlated variables

budget is decreasing. Such a condition might be relevant to identifying a root cause of cybercomplexity as discussed in Chap. 9.

11.5 A Stochastic Framework for Trust

Information entropy can be used to establish a framework for categorizing trust scenarios if a stochastic approach to identity authentication/trust management is adopted. As before, we assume trust management is a binary stochastic process, where the process outcomes are either a trusted or untrusted identity.

Therefore, there is a probability associated with each of the two outcomes, and the entropy values are once again identical to those shown in Fig. 6.9. As described in Sect. 11.3, the number of probable states (T) resulting from the identity authentication process can be estimated from the number of authentications in combination with the entropy per authentication.

An assignment of trust can be made based on this formulation, where each category reflects the magnitude of authentication identity uncertainty, which is of course specified by the entropy. To be clear, this framework is merely a taxonomy of scenarios based on the number of probable states resulting from a stochastic trust management/identity authentication process.

The value of identifying categories of trust is in recognizing the complexity associated with each category and the respective implications to security controls. This framework does not actually specify an absolute value of trust. Furthermore, it is not necessary to actually calculate the entropy associated with each category. The entropy merely serves as a reference or label that reflects the uncertainty inherent to identity authentication for a given scenario type.

The highest-trust scenario in this classification scheme is one that contains only one possible state since the uncertainty in identity authentication is zero. Hence the probability of successful authentication is unity. For convenience in labeling only we call this a Type One trust scenario. If there is indeed only one state, there can be no variability in the post-authentication states, which obviates the need for methods such as correlation.

A Type One scenario might include those where aural and visual inputs are used for identity authentication, which are assumed to be infallible when used in combination. Since there is only one possible state resulting from this type of authentication process the probability of the scenario being in any other state must be zero, which implies the identity authentication process entropy is zero. In summary, a Type One trust scenario is defined by zero entropy and hence a single probable state.

A Type Two trust category is a scenario where the number of probable states is greater than zero but less than the maximum. This condition can only occur if the trust management probabilities are not equal. The sole criterion for a Type Two trust scenario is a value of trust management/identity authentication entropy greater than zero and less than one.

Table 11.1 Trust management framework

	Type 1	Type 2	Type 3	Type 4
Entropy (H)	0	$0 < H < 1$	1	Undefined
Distribution of states	Single state	Number of probable states is greater than one but less than the maximum	Maximum number of probable states	Infinite number of states

Note once again it is not necessary to know the number of probable states resulting from the trust management process nor is it required to know which states are more or less probable. We merely need to know there is a spectrum of probable states resulting from the process.

The Type Three trust category is one where the number of probable states resulting from trust management is a maximum. Such a condition implies the entropy of the binary stochastic trust management process is one. Therefore, both outcome probabilities of that process equal 0.5. In other words, the trust management process is equivalent to a fair coin toss. Clearly, such a scenario is not ideal from a security risk management perspective, but there is an even more untrustworthy scenario, which is described next.

The Type Four trust category includes the lowest trust management/identity authentication scenarios. These scenarios are characterized by an infinite number of probable states. The operational implication is the probability of identifying any particular state is indeterminate. IT environments arguably possess an uncountable number of risk factors (and hence an exponentially large number of probable states) and therefore constitute a Type Four scenario, which is consistent with the historical difficulty in quantifying the potential for information compromise.

In summary, classifying categories of trust is facilitated by a stochastic approach to identity authentication/trust management. The categories derive from differences in the identity authentication process entropy. Table 11.1 specifies the four trust categories and the categorization criteria.

Chapter 12
Operational Implications

12.1 Introduction

A theory pertaining to cybersecurity risk is of questionable value unless it has at least a vague connection to the real world. Specifically, it should provide insight into the drivers of risk with respect to information compromise, which forms the conceptual basis for applying security controls to risk factors.

However, the theory must ultimately lead to methods that security practitioners can apply to actual IT environments. Merely knowing the root causes of cybercomplexity is operationally useless unless there is some hope of identifying them. Knowing when to apply security controls is as important as what controls to apply.

Equally, the limitations of the theory must be understood.

Successfully managing cybersecurity risk is the objective of every cybersecurity professional. Examining both individual and enterprise-level drivers of cybersecurity risk is necessary to developing a comprehensive cybersecurity risk management strategy.

12.2 Risk-Relevant Organizational Features

In Chap. 9 we discussed root causes of cybercomplexity. These are the progenitors of cybersecurity threat incidents, e.g., data breaches, hacks, etc. Ideally a cybersecurity risk management strategy will manage these root causes but they can be difficult to identify let alone address.

C. S. Young, *Cybercomplexity*, Advanced Sciences and Technologies for Security Applications, https://doi.org/10.1007/978-3-031-06994-9_12

Fortunately, certain risk-relevant organizational features are indicative of conditions that foment cybercomplexity. These features are generally easier to identify than the root causes themselves since they have well-defined characteristics. Such features are the subject of this section and are articulated immediately below.

Cultural dissonance within an organization portends badly for organizations attempting to reduce the potential for information compromise on an enterprise scale. Dissonant cultures can result in a lack of coherence and/or consistency in managing risk-relevant processes, workflows and technologies thereby increasing risk management uncertainty and/or the number of risk factors for information compromise.

Specifically, the various cultures or sub-cultures within an organization can be in tension, which can cause inconsistencies in the balance between business facilitation and security restriction. Therefore, the aforementioned processes, workflows and technologies can exhibit local deviations from the tolerance for risk across the enterprise.

Such deviations are more likely and grow in number as the scale of the IT environment increases. Organizations that have grown by acquisition are prone to cultural dissonance since each entity brings its own culture to the mix. However, this risk-relevant feature is certainly not limited to those organizations.

Another risk-relevant feature is repeated inconsistencies in implementing security controls across the enterprise. These can be a significant contributor to security risk management uncertainty, thereby driving the requirement for macroscopic security controls that promote uniformity.

This feature is highly scale-dependent, where a pattern of inconsistencies might not be apparent unless a sufficient segment of the environment is in view. Its effects might be equally transparent without a sufficiently macroscopic view.

Discontinuities and/or disconnects between organizational units represents a third organizational feature that suggests the presence of cybercomplexity. Numerous discontinuities create fertile ground for cybercomplexity since security risk management uncertainty is almost guaranteed. Such a condition is facilitated by a lack of communication and is also fueled by the absence of a cohesive force that can impose coherence on an enterprise scale, e.g., centralized security governance.

A pattern of delinquency in addressing security vulnerabilities also suggests the potential for enhanced cybercomplexity. In one common example, new security vulnerabilities are constantly being published and are subject to exploitation until remediation is achieved by patching. Keeping up with new vulnerabilities represents an ongoing struggle for many if not most IT departments.

The implication of delinquent vulnerability management relates to the overall approach to security risk management in addition to the presence of unaddressed vulnerabilities. If the rate of accumulation of high-severity vulnerability is significant it implies an increase in the number of risk factors. However, the point is the accumulation of such risk factors suggests more fundamental forces are at play such as an under resourced IT Department.

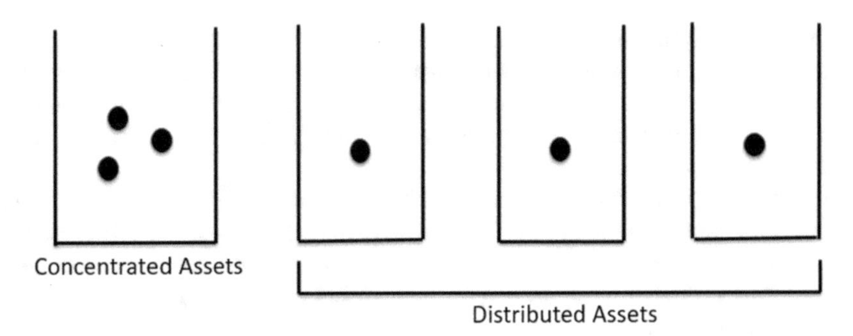

Fig. 12.1 Risk concentration versus risk distribution

"Risk concentration" is the uneven density of information assets. Such a condition immediately increases the impact component of risk due to the confluence of information assets in space and time. Per the definition of the impact component of risk, risk concentration results in an increase in the loss-per-incident.

Numerous instances of risk concentration would suggest insufficient security resources or perhaps a lack of security awareness. Either way, cybercomplexity is nearly assured if risk concentration is prevalent since all affected information assets immediately become risk factors for information compromise.

Figure 12.1 is a simple depiction of the concentration versus distribution of information assets.

A bureaucratic organizational structure implies the organization is stove-piped and/or overly hierarchical. Why is this organizational feature potentially indicative of cybercomplexity? A bureaucratically challenged organization is more likely to inhibit communication between business leaders and security risk managers. This condition increases the likelihood of security risk management inconsistency and incoherence on an enterprise scale.

A bureaucratic approach to management is actually a highly risk-relevant organizational feature. Fortunately, the signatures of a bureaucracy are often quite visible, and its deleterious effects can be partially offset by centralized security governance.

Incompleteness in security control implementation is another risk-relevant organizational feature. Although security technologies might be mandated on an enterprise scale it is not uncommon for unknown gaps to exist. For example, if anti-virus software is not distributed to all endpoints it immediately increases the number of risk factors for information compromise. Risk management uncertainty is also a by-product of incomplete security control implementation.

Incoherent security risk management is the final condition that suggests cybercomplexity is present in an IT environment. In this context incoherent means the "random" implementation and/or enforcement of security controls. The word random is in quotes because it has a rigorous meaning, which is intentionally ignored in the interest of making a point.

Incoherence in this context means the application of security controls to risk factors is untethered to a plan linked to a set of guiding principles, i.e., a strategy. The importance of reducing security risk management incoherence cannot be overstated, and a disproportionately lengthy discussion is therefore devoted to this topic.

The "Prime Directive" of cybersecurity risk management is to address the risk factors for information compromise and information-related business disruption via the application of security controls in accordance with an organization's tolerance for risk.[1] This mantra succinctly expresses the mission statement of every cybersecurity risk manager, which explains if not excuses its repetition *ad nauseam* throughout this text.

The tolerance for risk is affected by issues that are perpetually in tension. These issues include the impact of business processes that generate revenue juxtaposed with security controls that potentially restrict those processes. The financial and resource implications of security risk management efforts must also be incorporated into decisions on security control implementation in addition to their effect on business operations.

Coherent security risk management is defined as the consistent application of security controls to risk factors in accordance with the organizational tolerance for risk *on an enterprise scale*. As we have remarked in connection with other organizational features that presage cybercomplexity, evidence of incoherence and its effects become more consequential if not more apparent as the scale of the IT environment expands.

Establishing and maintaining coherence across a single business unit is typically easier than doing so across multiple units. Security controls implemented across increasingly expansive segments of the IT environment have greater opportunities to be misapplied, become ineffective or not be applied.

Maintaining security risk management coherence is a principal motivation for centralized security governance, a macroscopic security control discussed in Chap. 10. Centralization is essential since it helps reduce local deviations, where a pattern of such deviations on an enterprise scale would be emblematic of organizational incoherence and therefore a harbinger of cybercomplexity.

Incoherence can be manifest in many ways. For example, offices in different regions can implement varied security policies and/or adopt *ad hoc* practices seemingly without rhyme or reason. Incoherent risk management can also be manifest across multiple IT environment dimensions within the same business unit. Numerous

---

[1] In the TV and cinematic science fiction series Star Trek, the Prime Directive, also known as Starfleet General Order 1, is a guiding principle of Starfleet. It specifies that members are prohibited from interfering in the natural development of alien civilizations.

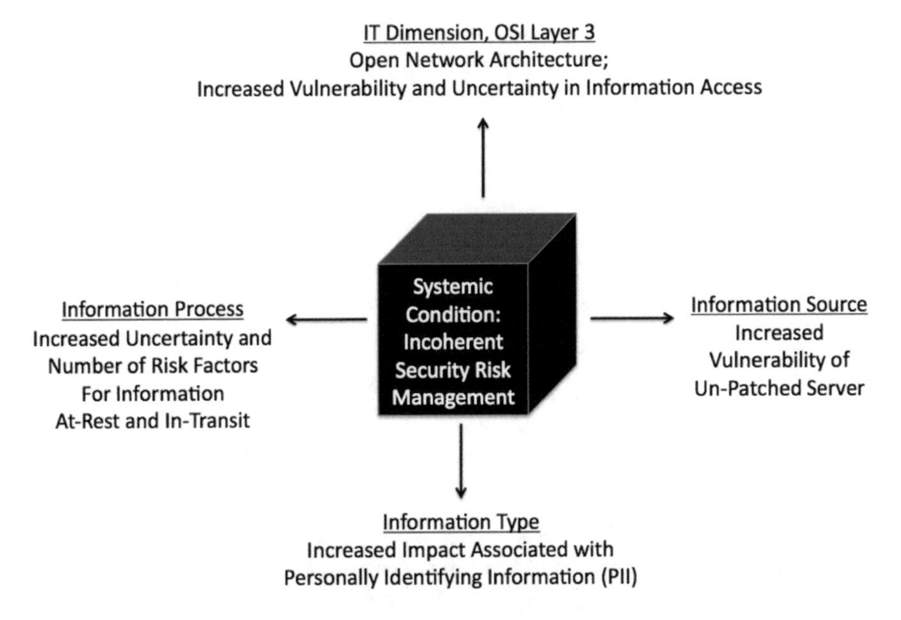

Fig. 12.2 The multi-dimensional impact of incoherent security risk management

unexplained deviations from the cybersecurity policy can be another warning sign of incoherence on an enterprise scale.

The effect of incoherence can be amplified within IT environments since each of the dimensions described in Chap. 4 does not exist in isolation. Vulnerabilities that are present within different dimensions can be related, and this relationship is sometimes exploited by attackers. Figure 12.2 illustrates this effect in action.

Departures from security best practice and/or deviations from the security policy could also result in an accretion of risk factors for information compromise. The magnitude of this effect will depend on their frequency of occurrence across the IT environment. Such deviations would be expected to increase risk management uncertainty.

It is possible to quantify the number of deviations if they are assumed to be a random variable. The identical calculation was performed in calculating the number of expected exceptions in IT environments.

Assume the deviation from security policy is a random variable x, occurs with a constant rate (λ) in a given IT environment, and the deviations are independent. The objective is to determine the probability of a specific number of deviations, which could provide evidence that the IT environment is ripe for cybercomplexity.

If the above conditions are true, the probability density p(x) of the expected number of deviations/departures is given by the Poisson probability distribution first discussed in Chap. 9,

$$p(x) = e^{-\lambda}\lambda^{x}/x! \tag{12.1}$$

For example, suppose λ corresponds to a constant mean rate of 2 deviations per risk-relevant process or workflow. If 5 such processes or workflows were analyzed, a total of $5 \times 2 = 10$ deviations would be expected. However, what is the probability there are *exactly* 5 deviations?

Equation (12.1) can be used to calculate this probability, where the exponential function $e = 2.72$, $\lambda = 10$ and $x = 5$.

$$\begin{aligned}
&\text{Probability of exactly 5 deviations}\\
&= e^{-10}\left(10^5\right)/5!\\
&= (e^{-10} \times 10^5)/120\\
&= \left(4.54 \times 10^{-5} \times 10^5\right)/120\\
&= 4.54/120\\
&\sim 0.04
\end{aligned}$$

Therefore, there is a four percent probability that exactly five deviations will occur in this scenario. Equation (12.1) can also be used to calculate the distribution of probabilities for a range of deviations. The result is shown in Fig. 12.3. The curve is not smooth because a small sample has been used in the calculation.

Equation (12.1) is a probability density, which means it specifies the probability with respect to a continuous variable (x). The total number of expected IT environment deviations can be calculated by integrating (12.1) with respect to x from zero to the number of relevant processes and workflows. The point of this exercise is to ultimately correlate the number of expected deviations with the potential for information compromise.

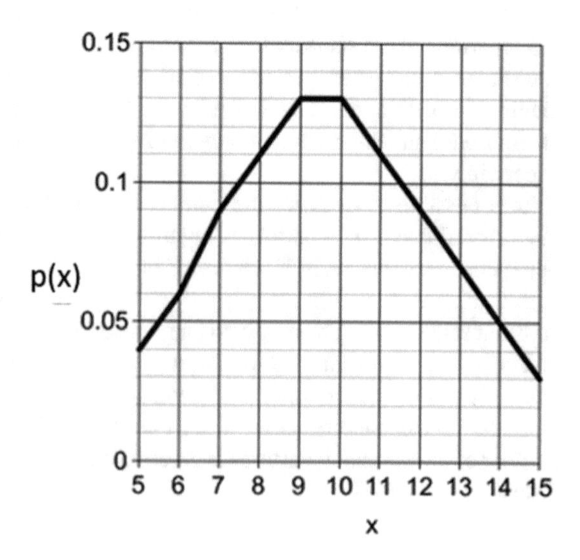

Fig. 12.3 Poisson probability distribution for $\lambda = 10$ and $5 \leq x \leq 15$

Unsurprisingly, IT environment incoherence is typically a by-product of the organizational culture. If the organization prizes convenience over security the environment is more likely to exhibit security risk management incoherence. In other words, incoherent security risk management and a convenience-centric security culture are potentially correlated.

To repeat for emphasis, incoherence is more likely to be present and/or observed as the time or distance scale of the assessment is expanded. Importantly, coherence is more difficult to maintain over larger spatial and/or temporal scales. Note if risk factors in IT environments were assumed to be a random variable whose values fluctuated in time, Fig. 3.4 would describe the devolution of the environment from coherence to incoherence.

12.3 Key Operational Results

The following is a delineation of the key operational results that arise from cybercomplexity.

(a) Modeling cybersecurity risk management as a binary stochastic process is driven by the need to ignore spatial and temporal fluctuations of risk-relevant conditions coupled with the large number of risk factors. The time scale of these fluctuations is typically short compared to the time scale of risk management efforts. The distance scale of fluctuations, both physical and virtual, is established by the distribution of information assets.

(b) IT environment complexity scales exponentially with the product of the number of risk factors and enterprise cybersecurity risk management uncertainty, where uncertainty is represented by the entropy of a binary stochastic cybersecurity risk management process. The unpredictability of the resulting states determines the magnitude of IT environment complexity, i.e., cybercomplexity. Cybercomplexity is the principal driver of cybersecurity risk on an enterprise scale.

(c) Even small changes in security risk management uncertainty can have a large effect on cybercomplexity especially if a large number of risk factors for information compromise are present. Therefore, even minor enhancements to risk-relevant organizational features can significantly affect the magnitude of cybercomplexity.

(d) The scale-dependence of cybercomplexity implies a broad perspective is essential to assessing the potential for information compromise across the enterprise, i.e., macroscopically. Only an appropriately broad perspective can encompass the full spectrum of risk factors, and a focus on individual vulnerabilities is not likely to identify and/or account for their full effect.

Conversely, a broad perspective will be too coarse to assess individual vulnerabilities. Severe vulnerabilities could also significantly contribute to the magnitude of cybersecurity risk depending on the type of vulnerability and the risk profile of the organization. A critical lesson of cybercomplexity is the scale of cybersecurity risk assessments must match the scale of the vulnerabilities within IT environments.

(e) Macroscopic security controls are effective in addressing cybercomplexity because they promote uniformity of risk-relevant processes, workflows, and technologies. The upshot is they reduce unpredictability on an enterprise scale. Non-uniformity across an IT environment complicates system administration, facilitates mishaps and/or encourages risk-relevant behavior by network users. Non-uniformity is manifest as unpredictability, which translates to cybercomplexity in this context.

(f) The requirement for uniformity applies to internal processes, workflows and technologies. In contrast, the environment presented to an attacker should be maximally complex to thwart efforts at random target selection. In general, complexity is an ally in the external/attacker frame of reference and an adversary in the internal/system administrator frame of reference.

Conflicting effects can result from certain security controls depending on the frame of reference. The underlying premise of reducing the so-called "attack surface" is that eliminating IT resources reduces the targets of opportunity of an attacker. For example, if an organization can support fewer browsers, the burden associated with vulnerability management is immediately lessened without undue consequence to network users.

(g) Although use of redundant information management methods enhances resilience, it can also increase one or more components of risk. An example is the use of forms to generate and store data. Uniformity in managing such forms via a single (and secure) solution decreases the likelihood and/or vulnerability to data leakage when compared with implementing multiple, *ad hoc* methods.

However, use of a single solution also concentrates risk. The key point is to ensure identified solutions utilize security controls in proportion to the potential for information compromise as assessed *in totality*, i.e., with respect to all three components of risk and cybercomplexity.

(h) Since every IT environment likely contains risk factors for information compromise, *all* cybersecurity threat scenarios exhibit cybercomplexity. Furthermore, it is not feasible to identify or enumerate all risk factors in an IT environment. Therefore, counting risk factors is a fruitless exercise, which in part motivates the statistical approach to characterizing cybersecurity risk on an enterprise scale.

Moreover, although information entropy quantifies security risk management uncertainty, it is impossible to precisely correlate the magnitude of uncertainty with the magnitude of risk especially with respect to an entire IT environment. Therefore, it is equally impossible to adjust security controls in order to fine tune cybercomplexity. There is nothing about the model for cybercomplexity that suggests it does not suffer from the same limitations as any cybersecurity risk assessment methodology.

(i) The operational implications of cybercomplexity can present a contradiction with attendant operational dilemmas. Namely, increased uniformity could actually *increase* the magnitude of cybersecurity risk despite a reduction in cybercomplexity. If every network communicator used a two-character password it would indeed increase uniformity and therefore reduce cybercomplexity. However, it would certainly not decrease the potential for information compromise!

Conversely, implementing certain security controls can simultaneously increase cybercomplexity but decrease the potential for information compromise. For example, network segmentation can increase cybercomplexity and simultaneously facilitate conformity with the Principle of Least Privilege. The lesson is every security control must be evaluated with respect to their effect on the entire IT environment and with respect to each component of risk.

12.4 Operational Limits

In Chap. 8 cybercomplexity metrics were presented that resulted from assuming cybersecurity risk management was a stochastic process. Although these metrics were accompanied by discussions on their applicability to real scenarios, each was acknowledged to have inherent limitations. The applicability of these metrics aside, how do revelations of cybercomplexity affect the security risk profile of a given IT environment?

Figure 12.4 is a "cartoon" that depicts the qualitative effect of macroscopic security controls on cybercomplexity through the imposition of uniformity on an enterprise scale. The net effect is to reduce the (hypothesized) mean value of risk management uncertainty as well as the "amplitude" of fluctuations about this mean value.

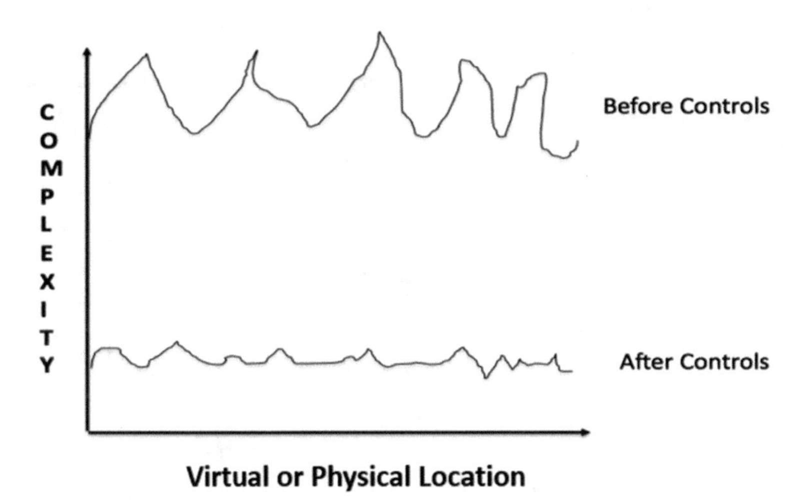

Fig. 12.4 Qualitative effect of macroscopic security controls on cybercomplexity

However, estimates of a mean value of security risk management uncertainty can vary significantly over short time and distance scales in IT environments. Representing risk management uncertainty via a single value represents a significant approximation, and is a limitation of this model of cybercomplexity and perhaps any model of a system or environment with large spatial or temporal fluctuations.

The operational impact of security controls addressing cybercomplexity will depend on the context and the particular component of risk affected by that security control. For example, the incremental advantage of creating subnets increases internal diversity if subnet access privilege is tied to identity.

However, from the attacker's point of view each subnet is equivalent since the attacker presumably lacks permission to access *any* subnet. Therefore, all subnets are equivalent to the attacker. Furthermore, increasing the number of subnets does not decrease the joint probability associated with the attacker identifying the correct asset. A simple example is illustrative.

Assume 100 information assets are uniformly distributed throughout an IT environment that has been partitioned via subnets. Only one information asset contains the information of interest to an attacker. The probability of selecting the asset of interest via random selection equals the probability of selecting the correct subnet times the probability of selecting the correct asset within that subnet.

Figure 12.5 plots the probability of selecting the correct asset as a function of the number of subnets. In each case the probability equals 0.01. Therefore, this security control does nothing to decrease the potential for information compromise via random selection in the attacker frame of reference.

Although diversity actually increased in the administrator's frame of reference, the potential advantages of information segregation conceivably outweighed the disadvantages that resulted from diversity. Such tradeoffs are common in security risk management, where the benefits of a particular security control can vary depending on the specific threat scenario.

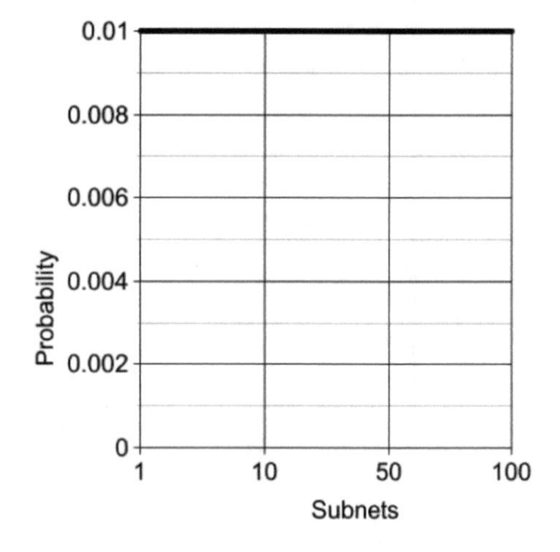

Fig. 12.5 The effect of subnets on the probability of random information asset selection

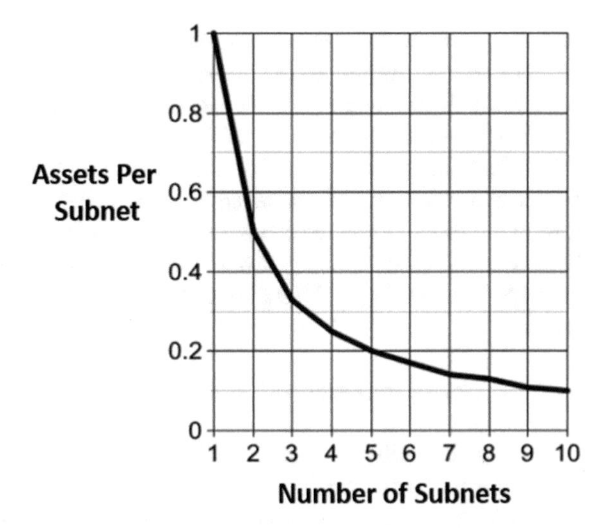

Fig. 12.6 Distribution of assets per subnet

Increasing the number of subnets does not affect asset targeting by random selection if the assets are uniformly distributed. However, the percentage of assets per subnet is not constant as the number of subnets increases. In other words, the *concentration* of assets varies, which affects the impact component of risk. Figure 12.6 plots this percentage for an IT environment consisting of 10 information assets and up to 10 subnets.

Managing cybersecurity risk inevitably demands tradeoffs. In this case the requirement to isolate information assets, and thereby enforce the principle of "need-to-know," must be balanced against increases to the impact component of risk arising from risk concentration.

Ideally the rate of decrease in the magnitude of the relevant component of risk resulting from security controls should be greater than the rate of increase in complexity that arises from these same controls. It is impossible to provide a general metric with more precision because the precise balance depends on the context.

12.5 The Potential for Information Compromise

The cybercomplexity model does more than identify a power law scaling relation for complexity in IT environments. That result in itself is significant because it explains why cybersecurity controls that promote uniformity should be a priority in any cybersecurity risk management strategy. The model also suggests a comprehensive cybersecurity risk assessment must consist of both macroscopic and microscopic security controls in order to fully address the likelihood component of risk.

The Fundamental Expression of Risk was first introduced in Chap. 1. Recall we focused considerable attention on the likelihood component of risk, and were careful to distinguish between the *potential* for a cybersecurity incident and the *probability* of same. Assessing the actual probability of a future security incident requires a probability distribution of similar historical incidents.

We also noted that such distributions in security risk management are rare. That said, a qualitative assessment of the likelihood component of risk is also risk-relevant, which means determining the potential for a successful attack. Considering what we now know about IT environments and particularly with respect to cybersecurity risk on an enterprise scale, it is worth revisiting the Fundamental Expression of Risk in the hope of achieving additional operational insights. Recall that expression was as follows,

$$\text{Risk(threat scenario)} = \text{Likelihood} \times \text{Vulnerability} \times \text{Impact}$$

The vulnerability and impact of information loss can certainly vary depending on whether the compromised asset is deemed "critical." However, in general the compromise of any confidential or even non-public information would be considered significant. Many such compromises would have serious implications to an organization's reputation and/or have legal/regulatory consequences.

Therefore, the focus of cybersecurity risk assessments is principally on the likelihood component of risk with the objective of reducing its magnitude to the extent possible. In other words, there is a general assumption that the vulnerability and impact components of cybersecurity risk are significant.

Since formulating a probability distribution of similar historical incidents is generally not possible, *assessing the likelihood component of risk implicitly becomes an assessment of the potential for information compromise.* Two assumptions apply in analyzing this potential in light of the cybercomplexity results discussed in previous chapters.

The first assumption is that IT Departments and Chief Information Security Officers are generally competent and well-intentioned. Therefore, any known gaps in security controls and discrepancies in the tolerance for risk would be addressed if these could be identified assuming the organization had the will and the means.

The second assumption is that any realistic IT environment contains numerous risk factors for information compromise and/or information-related business disruption. This claim has been a consistent refrain throughout this text. Given the multiple IT environment dimensions and the presence of risk factors within each dimension, an accretion of risk factors seems a near-certainty.

Since the principal driver of cybersecurity risk on an enterprise scale is cybercomplexity, the potential for information compromise must principally a function of two quantities: the number of risk factors and security risk management uncertainty. Therefore, we must understand the constituents of security risk management uncertainty.

The contributors to security risk management uncertainty consist of technical and non-technical sources. The technical source of uncertainty is gaps in security controls. The non-technical source is deviations from the organizational tolerance for risk. Presumably these gaps and deviations are not known. Therefore, the potential for information compromise, which recall is now a morphed version of the Fundamental Expression of Risk, can be expressed as follows,

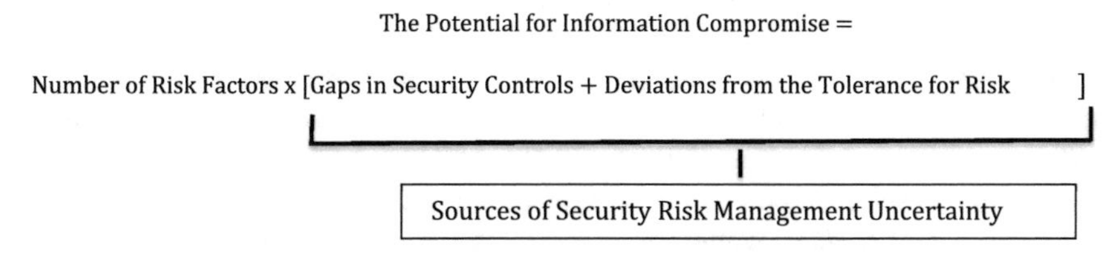

The key operational implication of this revised expression is that a comprehensive approach to addressing the potential for information compromise requires both microscopic and macroscopic cybersecurity risk management. Specifically, it requires focusing on the individual risk factors in conjunction with the technical and non-technical sources of security risk management uncertainty.

A microscopic approach includes identifying and assessing hundreds if not thousands of individual vulnerabilities. These result from automated vulnerability scans and require ongoing vigilance and attention. A macroscopic approach entails addressing the sources of security risk management uncertainty on an enterprise scale.

For example, deviations from the tolerance for risk include risk-relevant practices that differ from the published cybersecurity policy. Examples of excessive user permissiveness would represent a good start in identifying such deviations.

There are a number of references that specify the fundamental security controls. These include the NIST Cybersecurity Framework and the CIS Controls. These can be used to evaluate unknown gaps in security controls and thereby assess the technical source of security risk management uncertainty.

12.6 Cybercomplexity Assessments

It is natural to segue from a cybercomplexity model to a cybercomplexity assessment process or protocol. We are now aware that cybercomplexity is correlated with the presence of specific organizational features that spawn risk factors for information compromise and/or significantly increase risk management uncertainty.

Since counting every risk factor and accurately estimating risk management uncertainty is not practical, the presence of one or more organizational features is sufficient justification to assume cybercomplexity exists and thereby initiate a search for root causes.

Recall these organizational features relate to inconsistent, incoherent, delinquent and/or incomplete cybersecurity risk management processes. Their potential impact on information assets should be assessed, thereby revealing the potential for enhanced cybersecurity risk on an enterprise scale.

The interest is in identifying patterns of issues rather than the vulnerabilities themselves since a pattern is potentially indicative of a systemic issue. For example, an assessment of cybercomplexity would not assess the security of endpoints nor the quality of anti-virus software installed on those endpoints. Instead the focus would be on the completeness of security software installations across the universe of endpoints. The intent is to reveal missing installations thereby highlighting an accretion of unknown risk factors, which is evidence of risk management uncertainty and the presence of heretofore unknown risk factors.

In a similar vein, the persistence of critical vulnerabilities revealed during automated vulnerability scans would be of interest rather than the vulnerabilities themselves. Once such delinquencies are identified, an attempt would be made to identify the root cause of the accumulation and pursue remediation efforts in accordance with the tolerance for risk.

The collective impact of organizational features would be aggregated across all such processes and with respect to business units or some other relevant organizational component. The intent is to identify deviations from the organizational tolerance for security risk.

The vulnerability to specific threats would also be evaluated on an enterprise scale. For example, the likelihood and vulnerability to insider threats might be based on a review of the background investigation process in combination with identity and access management processes across business components and/or other relevant organizational units. In general, global inconsistencies, incoherence, incompleteness and dilatoriness that increase the potential for information compromise should be highlighted.

The high-level approach to managing cybercomplexity is summarized as follows:

- Identify organizational features that correlate with enhanced cybercomplexity.
- Assess the potential for information compromise and information-related business disruption based on the number of risk factors in conjunction with gaps in security controls and deviations from the organizational tolerance for risk across the enterprise.
- Identify root causes of cybercomplexity by identifying risk-relevant patterns on an enterprise scale.
- Apply relevant macroscopic security controls pursuant to addressing the root causes of cybercomplexity.

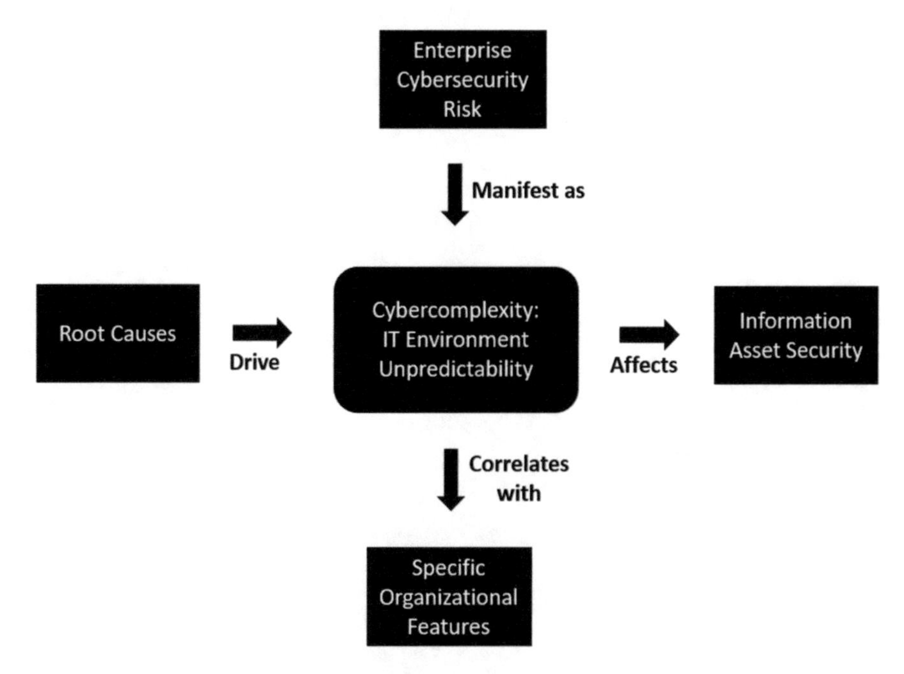

Fig. 12.7 Cybercomplexity assessment fundamentals

Figure 12.7 illustrates the fundamentals of cybercomplexity and its assessment within an IT environment.

Epilogue

Accepting the model of complexity as presented herein might be considered an acknowledgement of failure. After all, in this model security risk management is equivalent to a game of chance. Certainly games of chance can be fun and even instructive. But there is no avoiding the fact that security risk management is now mathematically identical to a coin toss.

One might counter by saying multi-faceted processes are neither entirely probabilistic nor completely deterministic. One could also argue that some uncertainty is associated with any such process, thereby opening the door to the relevance of probability distributions. Uncertainty is certainly no stranger to processes involving human behavior, which suggests probabilities are potentially applicable to security risk management. John Pierce articulates a similar view with respect to human behavior in general[1]:

Our experience indicates that the behavior of actual human beings is neither as determined as that of the economic man nor as simply random as the throw of a die or as the drawing of balls from a mixture of black and white balls. It is clear, however, that a deterministic model will not get us far in the consideration of human behavior, such as communication, while a random or statistical model might.

The real question is not whether a stochastic model of security risk management is an accurate representation of an IT environment. It surely is if one accepts a specific if narrow view of security risk management. The more relevant question is whether generalizing the results to more deterministic cybersecurity scenarios is valid. In other words, are the lessons derived from a stochastic model of security risk management applicable to IT environments that more closely resemble reality? The fact that the results agree with intuition is a good sign.

The motivation for invoking the laws of probability is to simplify IT environments by evaluating the risk factors *in aggregate* via statistics. The success of statistical mechanics in modeling physical systems has been a source of inspiration although there are clearly limits to the analogy.

---

[1] J. Pierce, op. cit.

C. S. Young, *Cybercomplexity*, Advanced Sciences and Technologies for Security Applications, https://doi.org/10.1007/978-3-031-06994-9

In the quest for simplicity only a binary stochastic model of risk management has been considered. Other models are certainly possible, and are arguably more realistic. For example, the mathematics of a die throw might be used to model security risk management rather than a coin toss, which would increase the possible outcomes from two to six.

Alternatively, cybersecurity risk could be viewed as a time series reflecting the combined effect of numerous, fluctuating risk factors. This is the approach taken in describing correlation-time in Chap. 3. The upshot is security risk measurement resembles a stock price and thereby exhibits so-called algorithmic complexity.

In truth, the simultaneous burdens and benefits of a stochastic approach overshadowed the details. Once the leap to randomness had been made, "the simpler the better" was a guiding principle in model selection.

One of the most significant results of this model of complexity is that cybersecurity risk is scale-dependent. The inescapable conclusion is truly comprehensive security risk assessments require engaging with the IT environments at the appropriate scale. It seems the model and intuition are in lockstep here since the world generally seems to operate this way.

Furthermore, it is the number of risk factors *combined* with security risk management uncertainty that drive complexity in this context. Critically, the effect is nonlinear, and the principal lesson is security controls that promote uniformity are required in order to reduce uncertainty on an enterprise scale.

Candidly, such conclusions are not profound. They simply argue for an expansive approach to security risk management that complements the one likely already in use. Drawing on yet another physical analogy, a microscope *and* a telescope are required to see microscopic and macroscopic phenomena, respectively.

Profound or not, traditional cybersecurity risk assessments tend to focus exclusively on individual vulnerabilities. Although this approach is frequently driven by necessity, the net effect is to potentially miss or underestimate risk-relevant effects that are only visible on an enterprise scale. This myopia might help explain the continued success of attackers in spite of the increasing reliance on technical security controls, noting the root causes of cybersecurity risk often have nothing to do with technology.

Ultimately cybersecurity suffers from its inherently defensive posture coupled with the inability to calibrate security controls. Unfortunately no one can predict what *would* have happened had certain security controls *not* been implemented. Nor can anyone determine how many security controls are necessary to reduce cybersecurity risk to some quantifiable threshold. In the absence of such thresholds strategy will invariably be overshadowed by tactics.

All commercial IT environments today exhibit complexity, which is an observation unlikely to raise many hackles. However, suggesting security risk management resembles a casino will undoubtedly inspire controversy. Yet it doesn't seem unreasonable to view probabilistic security risk management as a worst case. There is unlikely to be a more insecure IT environment than one where decisions are made with a pair of dice. Worst case results could be extrapolated to less extreme scenarios with appropriate caveats. At least that's the hope.

In addition, assessments of cybercomplexity are strictly about complexity and do not necessarily reflect the potential for information compromise. Macroscopic and microscopic cybersecurity risk factors are not necessarily correlated. Since a probabilistic approach is restricted to assessing complexity, security professionals need not feel unduly offended or threatened.

The reality is it might be impossible *not* to view complexity through a probabilistic lens if anything more than a superficial description of IT environments is required on an enterprise scale. Inevitably we might have to just accept the fact that uncertainty is inherent to cybersecurity risk when viewing the big picture. Perhaps the most discomfiting realization is that cybersecurity, like every other risk management discipline, is ultimately about identifying and managing the unknown.

Printed in the United States
by Baker & Taylor Publisher Services